Resilient Cities

Re-thinking Urban Transformation

Series Editors

Nicola Tollin, SDU Civil and Architectural Engineering, University of Southern Denmark, Odense, UK

Jordi Morató, UNESCO Chair on Sustainability, Technical University of Catalunya, Terrassa, Spain

Ernesto DR Santibanez Gonzalez, Centre Environmental Sciences, Federal University of Bahia, Bahia, Brazil

The RESILIENT CITIES book series aims to analyse the challenges faced by cities and provide an up-to-date body of knowledge, including a systematic collection of global cutting-edge best practices, fundamental to managing the urban transition toward resilience. The best practices will be collected and analysed following a common format, enabling the reader to understand the solutions adopted and clearly highlighting the parameters and possibilities for replication and up-scaling. The best practices are taken from a global city base including, Barcelona, Medellin, Adelaide, Copenhagen, Seoul, and Accra. The distinctiveness of the RESILIENT CITIES book series is its international dimension, coupled with a multidisciplinary and a cross sectorial approach. The RESILIENT CITIES book series will be a unique and fundamental resource for practitioners, policy makers and scientists involved in planning and governing the transition of cities. It presents the latest and up-to-date systematized information on research, practices and policies development, defining clear means and pathways for replication and up-scaling.

Jordi Morato · María Alejandra Taborda Caro ·
Luis Diaz · Massimo Leserri · Andrea di Filippo ·
Nicola Tollin · Carlos Arias · Lluis Garcia Petit
Editors

Intangible Cultural Heritage and Traditional Knowledge Related to Urban Resilience

UNIVERSITAT POLITÈCNICA DE CATALUNYA BARCELONATECH UNESCO Chair on Sustainability
unesco Chair
recnet.
recnet. RESURBE
Springer

Editors
Jordi Morato
UNESCO Chair on Sustainability at
Technical University of Catalunya
Terrassa, Barcelona, Spain

Luis Diaz
Technical University of Catalunya
Catalonia, Spain

Andrea di Filippo
Department of Civil Engineering
University of Salerno
Fisciano, Italy

Carlos Arias
Aarhus University
Aarhus, Denmark

María Alejandra Taborda Caro
University of Córdoba
Montería, Colombia

Massimo Leserri
Department of Civil, Environmental, Land,
Building and Chemical Engineering
Politecnico di Bari
Bari, Italy

Nicola Tollin
Department of Technology and Innovation,
SDU Civil and Architectural Engineering
University of Southern Denmark
Odense M, Denmark

Lluis Garcia Petit
Institute for Intangible Cultural Heritage
(IPACIM)
Barcelona, Spain

ISSN 2524-5988 ISSN 2524-5996 (electronic)
Resilient Cities
ISBN 978-3-032-12712-9 ISBN 978-3-032-12713-6 (eBook)
https://doi.org/10.1007/978-3-032-12713-6

This Springer imprint is published by the registered company Springer Nature Switzerland AG
The registered company address is: Gewerbestrasse 11, 6330 Cham, Switzerland

Foreword

Heritage often brings social connectivity. Social connection is frequently part of business models, especially for producers of consumer products with a clear sustainability ambition. To reach our goals in sustainability such as the UN 2030 SDGs (Sustainable Development Goals), transitions seem to be required and inevitable. Therefore, they are almost always mentioned in the same sentence with innovation. However, such transitions might be much easier achievable by the reuse of what is done in the past. How can we use our heritage to create a better more sustainable world that will allow our children to create their own heritage?

This is not about "Back to the future" but maybe more about "Our future is the past".

The industrial revolution in the second half of the 19th century is an example of a systematic transition. This industrialization initiated a different way of working and living, on average prosperity increased, but it also caused a large social and economic inequity which led to poverty under the working class. Big transitions seem to easily lead to inequity. We might want to look for ways to have necessary transitions improving quality of life, but avoid the effect of increasing differences between have and have nots.

The present day transitions we are facing are driven by changes in demography, consumer demand and climate, and are taking place in the energy, food and health sector. Similarly to the industrial revolution, most of these transitions tend to increase inequity

At IHE Delft we work on the development of capacity in the water sector as well as for water users. All of the before mentioned transitions have a huge effect on our water resources. Accessibility to water will lead to large differences in income, health and well-being. To steer these developments to a better, more sustainable future for all, we need among others to renew our relationship with water. Our culture reflects our relation with water. Water museums, in its broad context, can play a big roll to renew our bond with water, especially for our children who are our future. Industrial heritage can show the great things mankind is capable of doing. It helps us to remind us of an earlier transition changing our society. Are there lessons learnt, we can use to adapt, become more resilient and prevent social and economic injustice?

The chapters in this book will take you on a journey of case studies and traditional knowledge identifying socioecological resiliency. Similar to the water museums it shows through our cultural heritage the resiliency and adaptive capacity we have. Some examples and exhibitions may trigger your thinking and help build our future on lessons from the past.

I wish you a lot of pleasure reading this book and hope to welcome you and your (grand) children in one of our water museums in the World.

Eddy Moors
Rector IHE Delft Institute for Water Education
Chair Global Water Museum Network (WaMuNet)
Westvest 7, 2611 AX Delft
The Netherlands

Preface

Dear reader, the book you have in your hands, entitled *Intangible Cultural Heritage and Traditional Knowledge Related to Urban Resilience*, is part of the Springer collection *Resilient Cities: Re-thinking Urban Transformation*, which includes the initiatives of the Resurbe program, which since 2014 has organized several international meetings (Spain, 2014; Colombia, 2015; Mexico, 2016 and Argentina 2016) with the intention of sharing and systematizing experiences and reflecting on the field of urban resilience, with a view to participating in the III Conference of the United Nations on housing and urban development sustainability, HABITAT III that took place in Quito, Ecuador, from October 17 to 20, 2016.

Overpopulation, growing urbanization, impoverishment and environmental deterioration represent current trends that we cannot fail to consider when we imagine the future of cities and with them, of humanity (since 2007, more than half of the world's population lives in cities). At the same time, cities present varying degrees of vulnerability to natural events (hurricanes, earthquakes, tsunamis, pandemics, among others) and human events (accidents and terrorism) that increasingly threaten their stability, as well as the lives of their inhabitants and the environment.

We are living a stage of history that Ulrich Beck (1986) identified as 'The risk society'. How can we prepare for risk scenarios in which natural and human phenomena impact socio-ecosystems with increasing frequency and with increasingly destructive potential? With the firm purpose of joining the effort to approximate workable answers to these questions, this book brings together various contributions that include conceptual reflections and practical experiences (casuistic) around the contribution of intangible cultural heritage and traditional knowledge to urban resilience.

Concerning the need to sustain environmental functions and processes, today's society has needed commitment and interest in reinventing symbolic cultural practices and representations. Hence, sustainability is also understood as a human attitude to the ecological crisis and its cultural possibilities. In fact, the social capacity to transform the lifestyles of the contemporary world requires a multidimensional and socio-ecological conception of human welfare: a good life within the biophysical limits of ecosystems.

The relationship between heritage, resilience and sustainability is a reflective exercise through which the conservation of heritage can be understood. Whatever its tangible and intangible manifestation is, the conservation of heritage is fundamentally a practice of memory, of cultural legacies generated through time which have left lessons on how to defeat adversity through what has been collectively applied to overcome the shortage or excess demand for ecosystem services. Nowadays, the task is to work with the complex and dynamic relationships that are established between ecosystems and socio-cultural systems.

The complexity and multidimensional character that characterizes the phenomenon of urban resilience determine that its approach requires incorporating the perspective of its various actors and relevant disciplinary fields. In the two volumes that make up this book, the experiences of 56 authors are presented; each of them develops their work as part of a team in 32 different institutions, located in 15 countries (Argentina, Australia, Brazil, Colombia, Denmark, Germany, Greece, Italy, Ireland, Mexico, the Netherlands, Portugal, Spain, the United Kingdom and the United States of America), from 3 continents: America, Europe and Oceania. Their professional fields cover a wide range of disciplines from: (a) the natural sciences (agronomy, biochemistry, biotechnology, chemistry, ecology and environment, geography, human ecology, physics), (b) the social sciences and humanities (anthropology, communication, economics, sociology, history), (c) disciplines of design and engineering (architecture, artificial intelligence, chemistry, civil, computing, design, electronics, environmental, industrial, landscaping, spatial planning, public policy, territorial development, structural, systems, urban planning, telecommunications, transportation), as well as various fields of mathematics, applied statistics and operational research.

This book, based on the previous reflections, is divided into six topics. The first one focuses on opening the discussion of the dimension of the **value of cultural diversity as an achievement for humanity**. Cultural practices have a great influence on the relationship between humans and their natural environment, and the way natural resources are used, seen in various examples throughout the book, reveal the importance of cultural practices, social and environmental dynamics towards resilience and sustainability.

The second topic refers to an exercise of reflection between the concepts of **community resilience and biocultural heritage**: encounters and disagreements with environmental concerns. Articles show different points of analysis and theoretical changes of the environmental discourses; the passage of the green revolution to sustainable development, in which they attempt to understand ecological, social and communitarian resilience, under the relation capacity/adaptation of the anthropo-natural systems.

Topic three, called **socio-cultural adaptation**, begins with the authors examining the inter-relationships between situated knowledge and resilience-resistance. It is an exercise of interculturality and intertextuality. Within the studies of water adaptation presented here, we find the adaptive model based on the pre-Hispanic Zenú culture (La Mojana, Colombia), who with great engineering skills, achieved the construction of the first hydraulic system in the American continent. This region of Latin America has been home to the largest water complex in Colombia and Latin America, where a resilient and sustainable culture has been created, as evidenced by the Colombian sociologist Orlando Fals Borda in the '70s, and which today continues to be researched in order to reconsider its processes of change. Composed of about 500,000 hectares of surface, corresponding to housing units, agricultural areas, fishing, access routes, alternate routes of circulation and the formation of complex 'cacicazgos'. It is undoubtedly considered one of the greatest modifications of the pre-Hispanic natural landscape in Colombia. It includes a series of ridges and channels, of great complexity and interconnected with the major bodies of water (ciénegas) and 'esotos' in turn, allowing the flow of fresh water to salt water, allowing it to oxygenate with its freshwater–saline life of estuarine ecosystems or mangroves. These bodies of water are connected to rivers, establishing a water–land communication relationship that serves for their daily renewal and for the configuration of human–amphibian life. Life in this part of the world is perceived as resistant to variations in water levels. With this we can show a success story about the use and exploitation of wetlands, based on Nature-Based Solutions (NBS) used by the Zenú natives more than 500 years ago.

Topic four, **Culture, sustainability, and appropriate technologies**, aims to highlight the role of simple and intensive knowledge technologies, developed in Latin American and European countries, applying the ancestral knowledge of vulnerable communities in regions with extreme conditions for the supply of drinking water such as the Brazilian semi-arid area. Integrated and sustainable water management requires environmentally friendly technologies to treat water that is inevitably contaminated by daily human activities. For small urban and rural communities (generally under 2000 inhabitants), green technologies, which emulate natural processes and use a minimum of energy, contribute to their empowerment by allowing the reuse of water in different activities and the increase of economic opportunities.

Additionally, we see some culture-based adaptation experiences related to the use of constructed wetland treatment systems (CW) to mitigate surface water pollution in Mexico (Mx) and Colombia (Co). Likewise, research, carried out in vulnerable areas such as subtropical island archipelagos, located in the Atlantic Ocean, with one of the highest population densities in the world, shows the highest human resilience. Consequently, for more than 350 years, rainwater harvesting has been practiced in Caribbean islands such as Bermuda. It continues to be the main source of drinking water and the key feature of the architecture of vernacular buildings on the Caribbean islands. Bermuda is an island that has no rivers, streams, lakes or freshwater pools, but where rainwater has always been collected.

Also explored is how the process of seawater desalination becomes an option for large urban areas, where the extremely high level of demand allows production costs to be reduced to viable levels. Worldwide, an optimal technology for small communities and/or decentralized production has been developed that desalinates seawater using very little conventional energy and, instead of producing brine, produces commercial salt. The system is based on a large pyramidal structure (2500 m^2) that collects and reuses several times the energy of the sun.

The fifth topic, **cultural heritage, plants, and food**, gives us examples such as the southern Mexican states of Chiapas and Oaxaca, which possess the greatest wealth and natural/cultural potential, apparent in textile art, this potential is threatening to disappear. We find people of science, dedicating all their efforts to the dream of taking the communities out of the historical pain of poverty, by means of dignifying the work again. This work is essential to generate a model of life-based on the development of a sustainable agricultural system that produces inputs that will be used for the production of artisan products made of 100% cotton.

Countries like Mexico consume an average of about 30 million tons of corn for all of their different needs, of which more than 50% are imported. Of the total consumption of corn, about 10 million tons are used for food consumption in the form of tortillas, a staple of the national diet. About 7.5 million hectares are used for corn production in Mexico, and 80% of the crops use transgenic seeds. In Oaxaca specifically, there are 208,500 production units mainly for self-consumption. These areas are considered highly vulnerable to climate change. In the coming years, it will be a challenge to link peasant production with production markets, and therefore guarantee the life of rural and agricultural populations. Especially considering it a priority to consolidate their production and life systems without agrochemicals and developing relations of exchange of surplus with urban markets.

Within the studies of gender and heritage, the text presents the relationship of women with ethnobotany, mainly in the area of health promotion, diseases and feeding processes within the household, with the use of medicinal plants and vegetable resources from kitchen gardens.

From the above, within the sciences of sustainability, arises an interest to document aspects of Intangible Cultural Heritage (ICH), associated with health prevention, sacred beverages, home remedies and gastronomy, where one of the pioneer countries is Mexico, particularly the state of Oaxaca. It is important to highlight the importance of traditional cultures in the conservation of biodiversity and traditional ecological knowledge, which is a factor to be considered in the sustainable management of natural resources. The traditional drinks and rituals of their elaboration, as is the case of the Siaab g'ez, for the native peoples of Latin America, represent a manifestation of Intangible Cultural Heritage (ICH) in danger of disappearing and its possible impact on the biodiversity of the cocoa plantations and other native species.

We then find that from the studies of gender-focused on health, it is possible to contribute to a better understanding of the resilience of urban communities in the treatment of health. The social role of traditional healers and herbalists, and in the case of women shamans, are involved in the treatment of children's health with plants, with a long tradition in the field of alternative sciences and the aetiology of some diseases that are treated, such as parasites, joint pain, back pain, sprains, diarrhoea and cough.

If we talk about sustainability in relation to land cover and land use, we find in this volume an example that allows us to evaluate the changes in the surface of the earth generated by human activities. From the methodology of multi-temporal analysis, an intersection between the interpreted covers, to quantify the changes in the study period through comparisons and geographical analysis of the results obtained in the processes with each of the satellite images. The methodology allows for the identification of areas where changes in coverage and land use have occurred. These changes reveal tensions between cultural forms of the appropriation of nature and the disposition of the communities' food pantry.

The book also includes studies on factors affecting the food security of Emberá Katíos Indigenous families in Alto Sinú, Córdoba-Colombia. The adaptive resilience of the Native American people is due to the current dispute over land and its permanence, without the interference of the systems of the extractive economy. Food Security, referring to cultural resistance and environmental sustainability, represents the basis for the projection of life and the safeguarding of ancestral knowledge linked to water and forest ecosystems in the Colombian Caribbean.

In early human occupations, we present in the Altiplano (high plains) of the Andes Mountains in Peru, the Totorales de Huanchaco, whose area of influence is in the north of the South American country, especially the seaside resort of Huanchaco. The rescue of the traditions of the use of the *totora* (cane plant) as a millenary cultural element of the Chimú civilization, is an example of the importance of this plant as an essential raw material for the construction of the *caballito de totora* (little reed horses), whose cultural symbolism goes beyond national borders. It shows the development of handicrafts and tourism in the northern region, as well as the development of ethnobotany, and the vulnerability situation found in these natural ecosystems.

Finally, the sixth topic, is focused on **cultural heritage linked to traditional housing** and sustainability. An international journey from the Mayan Indigenous populations, all the way to India, and Turkey, shows us the new model of vernacular and participatory architectural construction applied in developing countries by international NGOs.

The Mayans of the Yucatan Peninsula, Mexico, have manufactured and improved their domestic architecture for over 3000 years. Pre-Hispanic evidence of construction is found at many archaeological sites. The Mayan house of wood and earth is an example of traditional knowledge being well adapted to the environment. The constructions are hurricane shelters, bioclimatic and sustainable, the secret of this practice lies in the selection of wood for the structure.

Another proven success story comes from India, where a 7.6 magnitude earthquake struck Bhuj, Gujarat, India, in January 2001, causing a large loss of life. Approximately 14,000 people were killed and 167,000 injured and property of about $3.2 billion was destroyed. The key challenge faced in the aftermath of the disaster was to provide affordable housing that is socially acceptable, of quality construction and structurally sound. The city of Bhuj responded with the construction of traditional Bhunga houses. Adapted earthquake-resistant construction methods and techniques, built from materials such as compressed stabilized earth blocks, thatched roofs, clay tiles and fly ash materials, from traditional houses that maintain the vernacular wisdom of the place.

In this line of thought, the experimental Bronze Age reconstruction, carried out within the Arslantepe archaeological land site in the Malatya Plain (Turkey), shows at the entrance of the museum, how a faithful architectural reconstruction to scale of an Early Bronze Age housing unit was erected to give visitors the opportunity to see the original architecture in its full form. The remaining walls of the excavated domestic structures are less than 0.5 m high, making it difficult for non-archaeologists to visualize the design and nature of the original community. This replica provides a physical sense of the individual and social space in the community. The aim is to highlight the importance of preserving and enhancing the tradition of earthen construction, as well as its associated knowledge and the set of skills transmitted from generation to generation, in order to avoid the risk of losing this intangible cultural heritage.

Concerning urban resilience, we have a case study of the historic Centre of Sardinia, Italy. This study focuses on issues related to the technological requalification of historical centres, the use of local stone material and the use of innovative energy efficiency techniques in line with the most innovative intelligent policies and paying specific attention to the landscape. Another study carried out with the Emberá Katíos Indigenous population of Alto Sinú in Colombia documents significant aspects of the community's architectural and cultural heritage, which they have called the children of water. Many Indigenous reserves in the country have begun to recover their own memory and historical heritage with the cooperation of local universities. This has allowed for the precise study of the architectural reality of the Tambos (called 'de' in the Êmbêra language).

As a characteristic element of areas with low rainfall, rainwater management in the archipelago of San Andrés, Providencia and Santa Catalina shows particular characteristics associated with the rainy season in the area. Historical, social and cultural aspects reflected in the constituent elements of the vernacular housing. These are manifestations of local knowledge, developed in contact not only with the environment and traditional elements transmitted from generation to generation but also with incorporated elements, since it is a system open to innovations and new technologies.

We are grateful to all the authors for their dedication, commitment and motivation to share their experiences. Without their devoted efforts, this book would not have come to fruition. After a review of a relevant demonstration of cultural diversity, traditional knowledge and appropriate methodology for the study and the identification of socioecological resiliency, we have to express deep respect for all those who throughout the centuries have shown the sacred value of nature. The human capacity for adaptation and resilience to change regards the search for establishing a spiritual connection and respect for nature, the living environment which provides home, food and shelter for thousands of living beings.

The human tool of culture is responsible for perpetuating our lives as we know it today. Through the process of resiliency, carried out in every corner of this planet, we can guarantee in a greater way the safeguarding of the immaterial cultural patrimony.

Life is tied to the human capacity of adaptation, and we do it with material and immaterial tools provided by culture.

Terrassa, Spain — Jordi Morato
Montería, Colombia — María Alejandra Taborda Caro
Catalonia, Spain — Luis Diaz
Bari, Italy — Massimo Leserri
Fisciano, Italy — Andrea di Filippo
Odense M, Denmark — Nicola Tollin
Aarhus, Denmark — Carlos Arias
Barcelona, Spain — Lluis Garcia Petit

Introduction

The concept of culture, besides being polysemic, is prodigious in the multiple interpretations it inspires. Never before in history have so many documents, research, essays and theories been written about contemporary culture. The fact is that the idea of this concept as it has traditionally been consolidated from a linguistic etymology, which is to say from its word, is giving way to another conception of greater nuances.

In addition, Burkert (1992, 2004) points out how culture was shaped from what was called 'the Greek miracle', a millenary philosophical thought in essence totalitarian and with little capacity to accept its hybridization and connection with the cultures of the Near East: Phoenicians, Semites, Acadians, Sumerians, Iranians and Egyptians. Our cultural heritage was shaped by a strong tradition to the oblivion of a shared past. Because of this, the Greeks were by excellence passive receivers of the ideas of others, calling those who did not master their language barbaric, and so they did not learn others either. On the other hand, Arendt (1996) describes how, conversely, the Romans were the first to have a better attitude towards the past and its meaning, seeking 'the sphinxes of the past', so they were the first to 'take culture seriously'. They felt they were heirs and continuers of the best of the past, they admired the achievements of others.

This is how the idea of culture was harmonized with that of civilization in the nineteenth century and well into the twentieth, so it was considered as a reference to designate some peoples more developed than others, catalogued as less advanced and even as barbarians or uncultured savages; the cultured came from the old continent. With liberalism, the struggle of civilization against barbarism was established, which would lead to the effective path of humanization; expressed in the domination of the man-nature (Martinell & Teixeira, 2011). These axioms have been transformed, as George Steiner showed in his 1971 essay 'In Bluebeard's Castle: Some Notes Towards the Redefinition of Culture', which stresses the idea of cultural evolution linked to progress and development spread from privileged centres.

These complexities have led to social sciences establishing a model or theoretical representation of this living reality that we call culture. The description and explanation of what is culturally accepted refer to a history, a space, a meaning and, within these, to endless connections that relate these cultural worlds with others, where the meetings are held not to delegitimize or deny but to reaffirm (Castro et al., 2010).

Today, for a good part of the academic tradition, culture is composed of a set of symbols, skills, ideas, values, statements, languages, attitudes and beliefs willing to be acquired through imitation, teaching and other forms of social learning, directly and indirectly through acceptance and recognition of certain actions or objects. Culture is information that is transmitted to condition the behaviour developed by individuals (Martinez, 2008).

Culture, in the opinion of one of the most illustrious anthropologists of the twentieth century, Clifford James Geertz, refers even more to a change in learned behaviour, but Geertz (1973) also relates this concept to a way of interpreting: *analysis of culture has to be, therefore, not as an experimental science in search of laws, but an interpretative science in search of meanings.* In a system of shared and conventional symbols created by man, the cultural elements behave as models historically transfer and materialize in symbolic forms, abstract and subjective rituals through which communities communicate, perpetuate and develop their knowledge about life and their attitudes towards it (Geertz, 1973).

Progressively, the semantic field of the word was extended, and with denominations such as interculturality, culture began to be conceived as a matter of 'incorporation': with rules of social conduct and forms of relationship with others and the world, which are slowly assimilated by each member of a community. Culture became the fundamental fabric of life in society (Stenou, 2002).

The Preamble of the UNESCO Universal Declaration on Cultural Diversity, (MONDIACULT, Mexico City), retained the definition of culture that was customary from 'arts and letters' until the early '80s (UNESCO, 1989, 2001). Later on, four stages were established in its evolution that the term culture has known since the last decades of the twentieth century and the beginning of the twenty-first. Maraña (2010) elaborated a synthesis of these that we present below.

In the '50s and '60s, culture was defined as an aesthetic and artistic expression. The new processes of decolonization in Africa and the trends of colonial equality, marked new discourses on the value of the cultural elements in politics and economy. Developing countries had demonstrated difficulties and little experience in protecting 'traditional' cultures. In addition, this moment was nourished by economic development theories, and orthodox Marxist ideologies, which identified culture and ethnicity as important for development, and generally supported policies of 'acculturation' that were notable in Latin America.

On the other hand, colonial struggles were no longer just political; national liberation movements vindicated national and ethnic cultures (Arispe, 2013)

Additionally, African UNESCO member countries began to pressure for new views and methodologies to classify the existence of their cultural manifestations, many of them worthy of being recognized as World Heritage Sites. Arispe (2013) describes the difficulties of this continent's expressions in getting its cultural wealth recognized, since they were based more on orality and other manifestations made with techniques and materials of short duration, unlike the untouched monuments and sacred sites in times when European cultures had them recognized as a memory of the world register. The dynamics of the permanent reconstruction of many African rites differ with the permanence and authenticity with which the cultural heritages are measured, from many of the other continents. It has been necessary, therefore, to place in dialogue multiple expressions, ways of life and present manifestations of the different human groups that share this planet.

During the '70s and '80s, culture and development began to be gradually linked. The role of the economy in this dynamic was defined, and so social cohesion, community integration and gender equality were established as public policies that integrated the issue of cultural diversity (Cevallos, 2005). The most important milestone at this point is the Convention concerning the Protection of the World Cultural and Natural Heritage, adopted by the UNESCO in 1972. The goal was to begin with the identification, protection and preservation of the world's cultural and natural heritage considered essential to humanity.

Between the '80s and '90s, a new turn began to take place with the integration of cultural issues linked to democracy, ethnic minorities, immigrant populations, cross-cultural relationships and the existing multiculturalism. Within the developed dynamics that countries were established, it was no longer necessary to grow economically solely, but rather to do so within a system where cultural, religious and political freedoms are realizeoreover, during the '90s and 2000s, the UNESCO Universal Declaration on Cultural Diversity was promulgated; adopted on November 2, 2001. This document insists on the value of the dialogue of cultures, the respect for all popular manifestations, and it gives a special place to the expressions of Indigenous groups, so that inclusion became an essential part of the common heritage of humanity. Today, UNESCO guides the understanding of culture as a continuous, broad, evolutionary process and defines cultural diversity as its main condition.

After several decades of academic and institutional discussions, the concept of development has an integral perspective; not only linked to high economic income per inhabitant, nor solely to a stable democracy and a strong system of values, instead it requires a triad: economic development, political development or democracy and cultural development, all at the same level.

Important Considerations on Cultural Heritage, Resilience and Sustainability

Culture, resilience and sustainability are concepts that today make up a triad that has not always been united, nor has it been worked upon in the same context. However, in recent years, new elements have been found along with other tools and ideas that promote increased attention to the integration of these concepts. For instance, nature can become part of both culture and heritage. As an example, the concept of 'cultural landscape' has recently been accepted, since the '90s. Undoubtedly, today nobody denies that culture and nature are in the same dynamic for the understanding of humankind and its survival (Heyd, 2006).

Cultural heritage as a denomination, must have been initially understood as a set of pasts and presents that define a community, such as its language, literature, music, traditions, crafts, fine arts, dance, gastronomy, clothing, religious manifestations, as well as its historical heritage. In the same way, natural heritage is configured with physical and biological formations, conservation areas with an exceptional universal, aesthetic or scientific value, places of habitats for animal and vegetable species, threatened or in danger of extinction, also as forms of expression, that is to say: literary, musical, plastic, scenic and recreational manifestations, among others (Protection of the World Cultural and Natural Heritage, 1972, UNESCO, 1972, Casas, 2016).

Tangible heritage is organized as a component of humanity as a whole, characterized by places and objects and customs with a strong moral, religious, artistic or historical component. These heritages are located in small groups, such as nations; or humanity as a whole, such as countries, departments and localities. Although the boundaries of these expressions are still under constant debate, their impact in relation to the survival of the various species that occupy the planet is no longer debatable. Ultimately, material heritage is a way of relating humans, nature and technology that is expressed through crystallized cultural value (SNPCN, 2018).

The treaties referring to the protection of cultural heritage come in different versions, tangible and intangible. They arise from the accelerated process of transformation and reinvention of new practices and cultural inventions, in the context of globalization and global environmental change, such as accelerated biodiversity loss, climate emergency and new habitat changes, which are transforming traditional cultural manifestations. The emergence of this concept and these orientations have required enormous discussions and complexities over time.

Territories require adequate cultural assessment that identifies their anthropological and natural connections, as well as the potential of the landscape resources, capable of developing feelings of collective self-esteem and appreciation for the uniqueness of their creative capacity. In the same way, collective memory is intimately related to the spread of cultures that are constantly changing, depending on the historical circumstances that allow them to do so.

Cultural heritage is beginning to be valued with the same ideas of sustainability; understood by Heyd (2006): 'As the set of resources of great potential, that have been transferred from the past to the present, the integrity of which must be protected and enjoyed and which can be increased, but not diminished or deteriorated before being passed on to future generations. From which the joint action of human beings and natural forces in the generation of certain heritage sites is directly recognized. Heritage is the determinant of valuable components of our world' (Heyd, 2006).

The concept of sustainability, as well as Ecosystem Services emerged in the '80s with the aim of analysing the impact of human activities on the environment. Since then, sustainability is related to the individual and/or collective capacity to satisfy current needs, without compromising the goods of future generations. In this sense, sustainability is linked to the conservationist actions of a society for the continuous enjoyment of ecosystem services, Fernández and Gutiérrez (2013) define it as 'the permanence of biological systems to remain diverse and productive over time, a necessary condition for the well-being of humans and other organisms'.

In this context, sustainability can be defined as a criterion of the possibility of long-term survival of natural and cultural elements, capable of accompanying any desirable human venture. Therefore, sustainability is the capacity to reproduce and revitalize essential human resources in the context of new forms of global market integration and new possibilities for intercultural dialogue. The concept of sustainability has been used until now, mainly in economic and environmental discourses on development.

At the same time that concepts of sustainability were being shaped and those of cultural heritage was being integrated into them, these were being strained by the economic views of development discourses, which did not integrate the relations between humans, nature and culture.

The category of resilience also emerged, within the social sciences from the '80s onwards, particularly incorporated by the field of psychology, to show people capable of remaining psychologically healthy despite living in high-risk contexts, such as poverty and multiple family issues, situations of chronic stress and internment canters.

On the other hand, the concept of community resilience has its emergence later, and can be applied to multiple social situations (earthquakes, floods, cyclones, famines, civil wars, guerrilla, repressions, etc.). These situations appropriate survival strategies, organizational schemes, commitment, linkage and other experiences to establish new social and labour bonds (Arciniega, 2013; Marzana, 2013).

Individual transformations affect collective influence, they aim to empower subjects and grassroots social organizations to change adverse realities with multi-causal views of the problems. Some theories have already begun to visualize this impact, such as Collective Efficacy developed by the economist Robert Sampson, who highlights the importance of an organized community as a potential for controlling the dynamics of security and coexistence of environments (Sampson, 1996).

Community resilience, as a collective circumstance to overcome and build upon disasters and massive circumstances of adversity, requires the organization of memory, of actions that over time have been designed as cultural practices, and empathic with the environment. This is the knowledge that enables adaptation to change, with creativity, for future orientation. One of the most important actions of this dimension is focused on the possibility of informing groups about the risks and emergency situations, with the aim of preparing preventive and repairing strategies

(Acinas, 2007). The collective allows for bonds of solidarity and learning with permanent meanings and symbols over time, with the idea of being able to act together and achieve expected effects.

Jordi Morato
María Alejandra Taborda Caro
Luis Diaz
Massimo Leserri
Andrea di Filippo
Nicola Tollin
Carlos Arias
Lluis Garcia Petit

References

Acinas, P. (2007). Información a la población en situaciones de emergencia y riesgo colectivo. *Intervención psicosocial: Revista sobre igualdad y calidad de vida, 16*(3), 303–322.

Arciniega. (2013). La perspectiva comunitaria de la resiliencia. *Psicología Política, 47*, 7–18.

Arendt, H. (1996) *"La crisis en la cultura: su significado político y social" en Entre pasado y futuro: ocho ejercicios sobre la reflexión política.* Ediciones Península.

Arizpe, L. (2013). Patrimonio cultural intangible: Los orígenes del concepto. En Lara, H. T., & Chávez, C. A. (Eds.), *Experiencias de salvaguardia del patrimonio cultural inmaterial.* Universidad Nacional Autónoma de México.

Beck, U. (1998). *La sociedad del riesgo. Hacia una nueva modernidad.* España.

Burkert, W. (1992). *The orientalizing revolution: Near eastern influence on Greek culture in the early archaic age.* Harvard University Press.

Burkert, W. (2004) *Babylon, memphis, persepolis: eastern contexts of Greek culture*, 192 pp. Harvard University Press.

Casas, N. (2016). *Patrimonio Mundial Cultural de la Humanidad.* España.

Castro, L., Nogueira, Y., & Toro, M. (2010). *Los origenes de la cultura humana ÉNDOXA: Series Filosóficas*, no. 24, pp. 219–245. UNED, Madrid. https://www.revistadelibros.com/articulos/evolucion-y-cultura-los-origenes-de-la-diversidad-cultural-humana

Cevallos, R. (2005). *¿Cultura y desarrollo? ¿desarrollo y cultura?: propuestas para un debate abierto.*

Fernández & Gutiérrez. (2013). Bienestar Social, Económico y Ambiental para las Presentes y Futuras Generacione). *Información Tecnológica, 24*(2), 121–130. https://doi.org/10.4067/S0718-07642013000200013

Geertz Clifford, J. (1973). *The interpretation of cultures: Selected Essays.* Basic Books.

Heyd, T. (2006) Naturaleza, cultura, y patrimonio natural: hacia una cultura de la naturaleza. *Ludus Vitalis, XIV*(25), 135–151. http://www.ludus-vitalis.org/ojs/index.php/ludus/article/view/492/494

Maraña, M. (2010). *Cultura y Desarrollo Evolución y perspectivas.* UNESCO http://www.unescoetxea.org/dokumentuak/Cultura_desarrollo.pdf

Marzana, D., Marta, E., & Mercuri, F. (2013). De la resiliencia individual a la resiliencia comunitaria. *Revista Electrónica Interuniversitaria De Formación Del Profesorado, 16*(3), 11–32. https://doi.org/10.6018/reifop.16.3.186961

Martinell Sempere, A., & Teixeira Coelho, J. (2011). *Políticas culturales: concepto, configuración y tendencia.* UOC.

Martínez García, B. (2008). El aprendizaje de la cultura y la cultura de aprender. *Convergencia, 15*(48), 287–307. Recuperado en 30 de diciembre de 2019, de http://www.scielo.org.mx/scielo.php?script=sci_arttext&pid=S1405-14352008000300011&lng=es&tlng=es
Ministerio cultura. (2010). *Eperara Síapidara.* http://observatorioetnicocecoin.org.co/cecoin/files/Caracterizaci%C3%B3n%20del%20pueblo%20Eperara%20Siapidara.pdf
Sampson, R. J., & Laub, J. H. (1996). *Socioeconomic achievement in the life course of disadvantaged men: Military service as a turning point, circa 1940–1965* (pp. 347–367). American Sociological Review.
Steiner, G. (1971). *n Bluebeard's Castle: Some Notes Towards the Redefinition of Culture.* Faber and Faber, UK.
SNPCN. (2018). *Sistema nacional de patrimonio cultural.* https://www.mincultura.gov.co/prensa/noticias/Documents/Patrimonio/Sistema%20Nacional%20de%20patrimonio%20cultural%202018.pdf
Stenou, K. (2002). *Declaración Universal sobre la Diversidad Cultural: una visión, una plataforma conceptual, un semillero de ideas, un paradigma nuevo.unesco.* https://unesdoc.unesco.org/ark:/48223/pf0000127162_spa
Beck, U. (1986). Der anthropologische Schock. *Merkur, 40*(450), 653–663.
UNESCO Organización de las Naciones Unidas para la Educación, la Ciencia y la Cultura. (1972). *Convention Concerning the Protection of the World Cultural and Natural Heritage.* UNESCO.
UNESCO Organización de las Naciones Unidas para la Educación, la Ciencia y la Cultura. (1989). *Recomendación sobre la Salvaguarda de la Cultura Tradicional y Popular*. UNESCO.
UNESCO Organización de las Naciones Unidas para la Educación, la Ciencia y la Cultura (2001). *Convención para la Protección del Patrimonio Mundial Subacuático.* UNESCO.

Contents

List of Figures

List of Tables

Chapter 1
Intangible Cultural Heritage as a Tool to Improving Urban Resilience

Lluís Garcia Petit

Abstract The concept of "intangible cultural heritage" (ICH), has been developed to safeguard cultural diversity, which is considered today as a boon for humanity. ICH enables the identification and valorisation of inherited knowledge and customary practices that have contributed to the resilience of our societies and that could continue to play this role or inspire new improvements. Therefore, ICH identification and safeguarding should be included in the strategies for sustainable development.

Keywords Intangible · Heritage · Sustainability

1.1 Cultural Heritage for Cultural Diversity

In the last few decades, the international community has realized that, unlike past ideologies supporting the idea of the superiority of some cultures, each culture is an achievement for humankind as it provides us with specific tools to survive as biological species. In its final declaration, the World Conference on Cultural Policies held in Mexico in 1982 stated among others[1]:

> 1. Every culture represents a unique and irreplaceable body of values since each people's traditions and forms of expression are its most effective means of demonstrating its presence in the world.
>
> 7. The international community considers it its duty to ensure that the cultural identity of each people is preserved and protected.
>
> 9. The equality and dignity of all cultures must be recognized, as must the right of each people and cultural community to affirm and preserve its cultural identity and have it respected by others.

[1] See the whole text at http://www.culturalrights.net/descargas/drets_culturals401.pdf.

L. Garcia Petit (✉)
Institute for Intangible Cultural Heritage, Barcelona, Spain
e-mail: lluis.garcia@ipacim.org

J. Morato et al. (eds.), *Intangible Cultural Heritage and Traditional Knowledge Related to Urban Resilience*, Resilient Cities, https://doi.org/10.1007/978-3-032-12713-6_1

> 10. Culture constitutes a fundamental dimension of the development process and helps to strengthen the independence, sovereignty and identity of nations. Growth has frequently been conceived in quantitative terms, without considering its necessary qualitative dimension, namely the satisfaction of man's spiritual and cultural aspirations. The aim of genuine development is the continuing well-being and fulfilment of each and every individual.

Following these reflections and by analogy with the concept of biological diversity, cultural diversity is seen today as a boon and an asset for humanity that we have to make efforts to preserve.

This idea was accompanied by the evolution and the enlargement of the concept of cultural heritage, which has been analysed by many researchers (see among others Llull, 2005, Ahmad, 2006, Loulanski, 2006, Martínez, 2006, Vecco, 2010, Selicato, 2016). Within the framework of the United Nations Educational, Scientific and Cultural Organization (UNESCO), founded in 1945, the first convention concerning cultural heritage, i.e. the Convention of The Hague for the Protection of Cultural Property in the Event of Armed Conflict (1954), does not provide a direct definition of the term, but declares that "cultural property" shall cover "movable or immovable property of great importance to the cultural heritage of every people, such as monuments of architecture, art or history, whether religious or secular; archaeological sites; groups of buildings which, as a whole, are of historical or artistic interest; works of art; manuscripts, books and other objects of artistic, historical or archaeological interest; as well as scientific collections and important collections of books or archives or of reproductions of the property defined above." So, in the praxis, we have a quite clear idea of how cultural heritage was meant at that time. Later on, the UNESCO Convention concerning the Protection of the World Cultural and Natural Heritage (1972) and the Recommendation concerning the Protection, at National Level, of the Cultural and Natural Heritage (1972) follow the same approach and they restrict cultural heritage to monuments, groups of buildings and sites.

In 1982, the above-mentioned World Conference on Cultural Policies became a milestone when it stated that: *"23. The cultural heritage of a people includes the works of its artists, architects, musicians, writers and scientists and also the work of anonymous artists, expressions of the people's spirituality, and the body of values that give meaning to life. It includes both tangible and intangible works through which the creativity of that people finds expression: languages, rites, beliefs, historic places and monuments, literature, works of art, archives and libraries."*

Even if this is not properly a definition of "cultural heritage", it shows a clear ambition for inclusiveness and embraces new conceptual fields, like intangible items. Despite this, some laws adopted in subsequent years continued to limit cultural heritage to material properties. This is the case of the Spanish Historical Heritage Act (1985)[2] and it was not until 2015 that the Act was modified to introduce a new paragraph stating that *"additionally, the properties which constitute the intangible cultural heritage are part of the Spanish Historical Heritage in conformity with their special legislation."*

---

[2] Ley 16/1985 del Patrimonio Histórico Español. https://www.boe.es/buscar/act.php?id=BOE-A-1985-12534&b=4&tn=1&p=19850629#aprimero

Since the late eighteenth century, the intangible aspects of culture and cultures were included under the concept of folklore. In the 1970's, the intensification of ethnological research in western countries led to the conceptualization of an "ethnological heritage", a term mainly used in France (Tornatore, 2004) and Spain, and not so much in other countries. As an example, the already mentioned Spanish Historical Heritage Act (1985) declares that *"the ethnological heritage is set up by the movable and immovable properties, as well as the knowledge and activities which are or have been relevant expressions of the traditional culture of the Spanish folk, in material, economic, social or spiritual aspects."*[3]

Some years later, UNESCO adopted in 1989 a Recommendation on the Safeguarding of Traditional Culture and Folklore, whose French and Spanish official versions avoid the use of the word "folklore" in favour of "popular": *Recommendation sur la sauvegarde de la culture traditionnelle et populaire* and *Recomendación sobre la salvaguardia de la cultura tradicional y popular*, respectively. In fact, paragraph A of the English text establishes the synonymy between the two terms: "folklore (or traditional and popular culture) is the totality of tradition-based creations of a cultural community…" In this way UNESCO introduced a new term for an already existing concept.

Despite the 1982 description of cultural heritage provided by the Conference held in Mexico, the idea of an "intangible heritage" was not formally assumed by UNESCO until the Proclamation of Masterpieces of the Oral and Intangible Heritage of Humanity, a programme launched in 2001. This was the last step before the adoption of the new concept of intangible cultural heritage established and defined in the 2003 Convention for the Safeguarding of Intangible Cultural Heritage. According to this text:

> The "intangible cultural heritage" means the practices, representations, expressions, knowledge, skills – as well as the instruments, objects, artefacts and cultural spaces associated therewith – that communities, groups and, in some cases, individuals recognize as part of their cultural heritage. This intangible cultural heritage, transmitted from generation to generation, is constantly recreated by communities and groups in response to their environment, their interaction with nature and their history, and provides them with a sense of identity and continuity, thus promoting respect for cultural diversity and human creativity. For the purposes of this Convention, consideration will be given solely to such intangible cultural heritage as is compatible with existing international human rights instruments, as well as with the requirements of mutual respect among communities, groups, and individuals, and of sustainable development.[4]

The Convention explains in the next paragraph that "this intangible cultural heritage is manifested *inter alia* in the following domains:

1. oral traditions and expressions;
2. performing arts;
3. social practices, rituals and festive events;

---

[3] Ley 16/1985 del Patrimonio Histórico Español. https://www.boe.es/buscar/act.php?id=BOE-A-1985-12534&b=4&tn=1&p=19850629#aprimero

[4] Article 2.1 of the Convention.

4. knowledge and practices concerning nature and the universe;
5. traditional craftmanship."

Compared with the previous existing concepts, two main differences can be observed. On the one hand, ICH is still alive. Extinct knowledge or practices are not considered within this new concept. It does not mean that they do not have any interest from the historical or scientific point of view, but they have not been transmitted to the present generations. The Convention focuses on preserving cultural diversity by avoiding the extinction of existing intangible cultural elements.

The second main difference lies in the responsibility to decide whether a given knowledge or practice is or not heritage, and the practitioners own sense of belonging. Until now, in practice, the assessment of the value or the interest of cultural properties or goods, and therefore their consideration as cultural heritage, was carried out by specialists, most often the specialists ascribed to the official institutions in charge of the implementation of the legislation related to cultural heritage. This approach has demonstrated a great effectiveness and has allowed for the protection of valuable heritage all over the world. However, the increasing number of properties, tourist exploitation and many other factors have developed, in many cases, a gap between cultural heritage and local communities. Within the framework of the 1972 UNESCO Convention, the need to involve the community in the protection of world heritage has been established in the last few decades and has become an increasing concern for competent authorities. In the case of intangible cultural heritage this is even more crucial, because it can only exist if there is a sufficient number of people ready to reproduce it.

This is a key issue that we should not overlook. The inertia of previous mechanisms is powerful, but we cannot afford to reproduce the system in which heritage is declared only based on the assumed expertise of a few individuals. The role of scientists, scholars, specialists, researchers or experts remains fundamental in order to provide the necessary information and knowledge about the uses, knowledge, practices and other manifestations that could be considered ICH. But only practitioners, either communities, groups or individuals, should decide if the element they keep alive is part of their cultural heritage. It is not always easy to make such a decision, among other reasons because concerned individuals are not always aware of the meaning of "intangible cultural heritage" or have not thought about the question whether a given practice provides them with a sense of identity and continuity. Therefore, a dialogue is needed between communities, experts, and institutions; information should flow in all directions with the aim that the community has all the necessary information to make a solid decision and official institutions acknowledge it. This is why it is so compelling to launch the identification process of ICH through inventories.

Beyond this, probably the most significant value of the new concept is that it has only one definition, generally accepted by the great majority of states in the world. Until December 2025, 185 states out of the 194 Member States of UNESCO had ratified the 2003 Convention and consequently accepted the definition of ICH included in it. None of the previous concepts benefited from such a consensus and this represents a major asset for the Convention and a solid foundation for the safeguarding of this kind of heritage. Consensus was finally reached after long discussions with scholars, specialists, governmental delegates, NGOs, community representatives and others.

1.2 Intangible Cultural Heritage for Sustainability

If we analyse the domains suggested by the Convention in which ICH manifests *inter alia*, we can find the most common fields associated to folklore or popular and traditional culture, i.e., legends, music, songs, dances, festivals, craftmanship, etc. At the same time, a less common category appears as "knowledge and practices concerning nature and the universe", crucial for the resilience related to environmental issues. The cultural values or aspects of the items included in this category have often been disregarded, probably because they are closely related to productive activities. On the other hand, it demands a greater involvement of specialists from this field, thus establishing new links between cultural and natural disciplines.

The consideration of this category is coherent with the principle stated in the introduction of the Convention, ICH is a guarantee of sustainable development. Indeed, cultural practices have a great influence on the relationship between humans and their natural environment, and the way we use the natural resources. We should not forget that sustainable development includes not only environmental aspects but has also cultural and social components. Today, we are more aware about the influence of these cultural and social components on the resilience of a society, and this volume is also thought of as a contribution in this sense, even if much remains to be done.

In this regard, the Representative List of Intangible Cultural Heritage of Humanity is a tool designed for the Convention to raise awareness of the significance of ICH and it includes many examples of elements that illustrate the way ICH contributes to resilience. Sega Tambour of Rodrigues Island (Mauritius) is a performance of music, song and dance which originated in slave communities. It facilitates conflict resolution, fosters socialization and consolidates bonds.[5] In Jeju Island (Republic of Korea), a biosphere reserve, there is a community of women who dive 10 m under the sea to gather shellfish, a practice that has contributed to the advancement of women's status in the community and promoted environmental sustainability with its eco-friendly methods and community involvement in the management of fishing practices.[6] Washoku, traditional dietary cultures of the Japanese, notably for the celebration of New Year (Japan), is a social practice based on a set of skills, knowledge, practice and traditions related to the production, processing, preparation and consumption of food, and it is associated with an essential spirit of respect for nature that is closely related to the sustainable use of natural resources.[7] The irrigator's

[5] https://ich.unesco.org/en/RL/sega-tambour-of-rodrigues-island-01257.

[6] https://ich.unesco.org/en/RL/culture-of-jeju-haenyeo-women-divers-01068.

[7] https://ich.unesco.org/en/RL/washoku-traditional-dietary-cultures-of-the-japanese-notably-for-the-celebration-of-new-year-00869.

tribunals of the Spanish Mediterranean Coast—the Council of Wise Men of the plain of Murcia and the Water Tribunal of the plain of Valencia—(Spain) are traditional law courts for water management, providing cohesion among local communities and a guarantee for a sustainable use of water.[8] The *oshituthi shomagongo*, marula fruit festival (Namibia), is a celebration lasting two to three days during which the members of the community, while preparing and consuming omagongo—a beverage made from marula fruit—they discuss family problems, sing songs, recite poems and share knowledge about basketry and pottery-making.[9]

Additionally, the List of Intangible Cultural Heritage in Need of Urgent Safeguarding includes some examples, like Hezhen Yimakan storytelling (China), a practice which, among others, preserves traditional knowledge of shamanic rituals, fishing and hunting.[10] The Register of Good Safeguarding Practices contains some examples of actions closely related to the resilience of the communities. The revitalization of the traditional craftmanship of lime-making in Morón de la Frontera (Seville, Andalusia, Spain), for example, is a source of employment and has extended the traditional use of lime in buildings.[11] Finally, the inventory of ICH in the Montseny Biosphere Reserve has provided new insights for the sustainable development of this territory[12] and allowed the local community to identify the following knowledge and practices concerning nature and the universe:

- Draught animals
- Apiculture
- Water and Forest Exploitation
- Breaking ground
- Farmers Calendar
- Charcoal production
- Grain and fodder cultivation
- Fruit trees cultivation and tree plantations
- Horticulture
- Grazing
- Cork bark extraction
- Cultivation of apple tree varieties from Montseny
- Popular weather forecasting
- Oil production
- Semi-stabling stockbreeding
- Popular medical cures and practices
- Moon phases follow-up

[8] https://ich.unesco.org/en/RL/irrigators-tribunals-of-the-spanish-mediterranean-coast-the-council-of-wise-men-of-the-plain-of-murcia-and-the-water-tribunal-of-the-plain-of-valencia-00171.

[9] https://ich.unesco.org/en/RL/oshituthi-shomagongo-marula-fruit-festival-01089.

[10] https://ich.unesco.org/en/USL/hezhen-yimakan-storytelling-00530.

[11] https://ich.unesco.org/en/BSP/revitalization-of-the-traditional-craftsmanship-of-lime-making-in-moron-de-la-frontera-seville-andalusia-00511.

[12] https://ich.unesco.org/en/BSP/methodology-for-inventorying-intangible-cultural-heritage-in-biosphere-reserves-the-experience-of-montseny-00648.

- Wine culture, crop and production.

1.3 Conclusions

All the mentioned items in the preceding paragraphs are examples of identified ICH elements that have an impact on the resilience of the respective human groups concerning environmental issues. In fact, all ICH elements, including oral expressions, festivals, songs or beliefs, have a role in the resilience of a community or society, not only because of the knowledge they embody, in addition they contribute to its cultural identity and social cohesion.

The cases presented in this volume provide new insights into the fundamental role that ICH has in nourishing or reinforcing the resilience of our societies. Even if not all of them have been formally inventoried as ICH, the message they transmit is that we should progress more synergically in the way towards the overall identification of ICH and the enhancement of urban resilience. In other words, ICH identification and safeguarding should be included in the strategies for sustainable development.

References

Ahmad, Y. (2006). The scope and definitions of heritage: From tangible to intangible. *International Journal of Heritage Studies 12.*

Llull, J. (2005). Evolución del concepto y de la significación social del patrimonio cultural, *Arte, Individuo y Sociedad 17.*

Loulanski, T. (2006). Revising the concept for cultural heritage: the argument for a functional approach. *International Journal of Cultural Property 13.*

Martínez, C. (2006). *El Patrimonio Cultural: los nuevos valores, tipos, finalidades y formas de organización*, doctoral thesis.

Prado, R., Taborda, A., & Cuello, Y. (2026). Emergence of communitarian resilience and biocultural heritage: interactions and conflicts related to environmental issues. In: J. Morató et al. (Eds.), *Contribution of Intangible Cultural Heritage and Traditional Knowledge to Urban Resilience.* Berlin: Springer. (RESILIENT CITIES: Re-thinking Urban Transformation Series).

Selicato, F. (2016). The Concept of Heritage. In F. Rotondo et al. (Ed.), *Cultural Territorial Systems: Landscape and Cultural Heritage as a Key to Sustainable and Local Development in Eastern Europe.* Springer.

Tornatore, J.-L. (2004). La difficile politisation du patrimoine ethnologique, *Terrain 42.*

Vecco, M. (2010). A definition of cultural heritage: from the tangible to the intangible. *Journal of Cultural Heritage 11*(3).

Chapter 2
Emergence of Communitarian Resilience and Biocultural Heritage: Interactions and Conflicts Related to Environmental Issues

Rafael Prado, Alejandra Taborda, and Yamith Cuello

Abstract Communitarian resilience is configured from a network of objective relations, between positions of power, knowledge and transformations over time. This work demonstrates in three moments, what is now called environment, as part of an academic, disciplinary and organized speech and how its construction path has reached what it is now called communitarian resilience, encompassing the green revolution and sustainable development. Initially, it will analyse the environmental science concepts' passage from the disciplinary to the transdisciplinary through the positioning of systemic thinking originating from the tensions of physics, chemistry and life sciences, dealing with terms such as: instability, non-linearity, crisis and relativity. Secondly, how the transdisciplinary view of science regarding the phenomenon of environmental deterioration can enable the creation of an environmental science concept which is capable of embracing the interdisciplinary studies in order to achieve the SDG. Considering the current transition of the resilience emergency and the changes that this speech has had from an ecological to social (communitarian) resilience within relation capacity/adaptation, contribute to conceptualizing environment science definitely in an inter and transdisciplinary perspective, albeit, initially causing tensions and disruption in social sciences. Thirdly, it will examine the place of a sustainable development speech in the face of communitarian resilience and how it could be presented from a study that is carried out in global conference speeches, i.e., United Nations; and finally, to consider the concept of resilience discussion which is connected and related to biocultural heritage from a political ecology perspective.

Keywords Communitarian resilience · Sustainable development · Transdisciplinary · Environmental science

R. Prado (✉)
UNESCO Chair on Sustainability—Polytechnic University of Catalonia, Barcelona, Spain
e-mail: pradojus@gmail.com

A. Taborda · Y. Cuello
University of Córdoba, Montería, Colombia

J. Morato et al. (eds.), *Intangible Cultural Heritage and Traditional Knowledge Related to Urban Resilience*, Resilient Cities, https://doi.org/10.1007/978-3-032-12713-6_2

2.1 Introduction

The emergence of communitarian resilience and biocultural heritage is the result of a positioning of environmental science environmental science and, in turn, of the grounding of the economic and political objectives of development. But the establishment of environmental studies could be analysed from different perspectives. On the one hand, those environmental studies were motivated by the tensions produced by the Theory of Relativity (1900) and the substitute proposal that involved the uncertainty principle, generating a change of paradigm: as of the nature pertaining to the rules and laws of nature so as to be understood as a multiple and mutable complexity. According to Prigogine (1996), for example, [since the Theory of Relativity] time is no longer the centre of attraction, which facilitated the change of immutable ideas, evolutions, crisis and instabilities.

On the other hand, this emerged from a speech made by the United Nations (UN), the UNESCO and the Organization for Economic Cooperation and Development (OECD) within the framework of general development planning, the natural environment, natural resources and, in addition, the green revolution, which transformed the food map of the planet from the 60's to the 80's of 20th Century.

Earlier speeches, after the Second World War, seemed to juxtapose themes related to sciences, but the fact is that these speeches promoted another theme: systemic thinking, which allowed for interdisciplinarity and, subsequently, transdisciplinary in a logical perspective *continuum* of the science. Such logic allowed us to understand how studies on the environment and the constitution of an environmental science were configured.

While these speeches are still being developed, at the end of the "short twentieth century" Hobsbawm (1999), the environmental speech is anchored in an economic and political perspective, encompassing keywords such as: development, sustainable development, resilience and biocultural heritage. Such movements are observed in the review of documents issued in the various summits and conferences held by the UN or international institutions attached to the UN.

In order to better understand the emergence of communitarian resilience, this Chapter is divided into three parts: (i) the first relates to the emergence of environmental science from a transdisciplinary view; (ii) the second part addresses the speech of interdisciplinarity as a practice for development; and, (iii) the third will focus on the passage from sustainable development to communitarian resilience which, as will be seen, needs to be articulated with the biocultural heritage from a political ecology perspective, given the necessity to deconstruct the category of resilience.

2.2 From the Disciplinary to the Transdisciplinary: The Place of Systemic Thinking and the Constitution of Environmental Science

The concepts of system, complexity and transdisciplinary, although semantically interdependent, structure themselves from changes in the scientific paradigm at the end of the 19th Century. The tensions placed between the unquestioning of space and time and physics were: the Theory of Relativity of Einstein (1905), the principle of exclusion of Pauli (1925), the principle of indeterminacy of uncertainty of Heisenberg (1925), the principle of complementarity of Niels Bohr (1949), and, consequently, classical physics founded on the logic that every physical phenomenon could be understood from a continuous chain of causes and effects. The previous changes gave another dimension to the disciplinary knowledge of sciences. In this regard Basarap (1995) expressed himself as follows:

"To a stranger, it may seem paradoxical that it is precisely from the very heart of the exact sciences that we reached into an idea of the limits of disciplinary knowledge. But from the inside, it provides evidence of the fact that, after a very long period of time, the disciplinary knowledge has reached its own limitations, leading to severe consequences not only for science, but also for cultural and social life" (p. 5).

The creator of the transdisciplinary study group from UNESCO (2002), Nicolescu Basarap, expounds on how concepts are built in classical physics, taking into account: continuity, local causality, determinism, objectivity, among others, and how they collide with the principle of discontinuity of Max Planck's quantum theory (1900), after carrying out his investigation on the problem of balance between radiation and matter (Otero, 1992). The result was published in 1906 under the title "Lessons on the Theory of Thermal Radiation".

Until then, physics had been trapped in the Newtonian dilemma and, in particular, with Atomistic Theory. Max Planck proposed instead that energy is a discrete quantity and non-*continuum* as it was considered back then. This was the beginning of quantum mechanics that enabled the comprehension of environmental phenomena on an atomic scale. In this regard, a new scientific paradigm was put forward which proved that the homogenous view of science was in crisis.

This approach would be continued by Albert Einstein's method, as well as the aforementioned authors and Lupasco, who would formulate the logical axiom of the "third included" or logic of inclusion, seeking to incorporate the "excluded third part", characteristic of classical logic, in an inclusive level of reality where contradictions coexist. According to Basarap (1995) the action of the logic of the "inclusive third part" over the different levels of reality, "induces an open structure […] of a set of levels of Reality. This structure has a considerable weight on the theory of knowledge, since it implies the impossibility of a complete theory, closed on itself" (p. 40–41). Such an approach will have a relevant impact on the theorization of transdisciplinary and, therefore, on the understanding of reality in a scientific perspective because it allows an "open unification" of the subjects/disciplines.

Faced with the reduction and isolation of the objects of study of the classical science in question from the advances already achieved in physics, it is required that complex totalities of the elements of the observed universe be studied under a relational logic; that is, in a systemic perspective not under a mechanistic and chaotic perspective that was dominant in the 19th Century. Such a perspective was not founded on a system but on objects in which its unit was reached by the sum of its parts. The new paradigm would establish a new landmark, "from an atomism centred on the constituent parts to an interactionism that gave rise to emerging systems" (Flórez et al., 2017, p. 28).

In this context, emerges the General Theory of Systems by Ludwig von Bertalanffy in the 30's, but only towards the end of the 60's were his ideas systematically published. This theoretical framework substantially affected the configuration of this new paradigm, by making one of the first approaches to the understanding of systems from the perspective of biology.

When this author states that "the feature of life does not reside in a distinction of singular processes of life [Lebensvorgänge], but of a certain order among all the processes" (Bertalanffy, 1934), it is establishing that order and organization, which derives from the whole, constitute the organism and, therefore, if it were observed separately would not reveal anything new about it; in order to understand that knowledge of the interdependencies of the subsystems that comprise it is necessary. In this view, biology needs to be approached from different methodological perspectives for the life system study, for example.

The notion of system is not recent, it has been present since classic times but subsumed in a closed unit as a set of interrelated parts. It was Luhmann and Morin who started to break with this classical logic, understanding the idea of open systems. "from the form of exchange of energy or matter with the environment, several types of systems have been established in physics: isolated, closed and open systems (Rumer and Ryvkin, 1980). In closed systems the exchanges of energy are produced, but no matter exchange with the surrounding environment, opposite to open systems, where there are exchanges both from energy and matter with the environment, and compensations like those that have place in living organisms". This contemporary approach of the system, as an open and live organism, makes possible the dynamism and the interrelations between elements and the system's interactions, taking into account the environment and the systems surrounding it and resulting in the unit as a product of the features of interaction.

In this direction, the speech sustained by the Theory of Systems proposes considering an existing interaction among organisms, energy and matter flows, originating from some interactions between biotic and abiotic factors which constitute an ecosystem. This concept was initially proposed by the English ecologist Arthur Tansley in 1935, later examined in depth by Lindeman (1942) and Odum (1953). According to Armentas et al. (2016), those authors were those "who began to discuss about the importance of energy transfer through the understanding of trophic chains" (p. 84). Even so, the conceptualization of the ecosystem has had different meanings: unity, self-contained identity, organization and system (Armentas et al., 2016).

Indeed, the discontinuity that has manifested itself in the quantum world and the logic of the inclusive third part has also manifested in the structure of reality, to the extent that it will be composed of different levels. Insofar as phenomena and nature are considered not being homogeneous anymore, because if local causality and interpretative mechanisms are questioned, this leads to the emergence and positioning of systemic thinking.

This new paradigm will allow us to understand the level of reality from a suprasystem- subsystem relation and not anymore juxtaposed (Martínez, 2011); that means, as a whole composed of different levels. In this theory "objects leave their place to systems; instead of essences and substances, organization; instead of aggregates that form a body, system of systems" (Morin, 1981, p. 148).

Understanding those levels of reality, "requires, then, different levels of logic to be applied", taking into account that the subjects/disciplines operate the knowledge of the parties, defined from a specific knowledge field that is limited to their object of study, when would be answered by a single level of reality. For this, it is necessary an interdisciplinary and transdisciplinary perspective that recognizes the disciplinary roots because they are not antagonistic, instead it helps towards an understanding of the phenomena. According to this, the three pillars of transdisciplinary are: the levels of reality, logic of the inclusive third part and complexity (Basarap, 1995).

The theorization of complexity arose in natural sciences in the late 1960s and in the early 1980s arose the concepts of emergency, self-organization and *autopoiesis*, as well as topics related to the origin of matter, life phenomenon, the relations between mind and brain and social systems, moving towards to the theoretical foundations that constitute, for example, the work of the sociologist Edgar Morin: *L'Esprit du temps* and *Le vif du sujet* (1992).

Edgar Morín seeks to propose an epistemology of complexity in *Le méthode*. The first part (out of six), published in 1977, was entitled *La nature de la Nature*. Morin seeks to articulate the science of man with the science of nature, in view of the fact that the great incision between them is due to the "omnipotence of a principle of disjunction". That is why knowledge of the most physical object cannot be dissociated from knowing the object rooted in a culture, in a society, or in a history. It is necessary to study all the physical knowledge in an anthropo-social perspective, as well as to study the social reality in its physical roots (Morin, 1981).

These relationships are configured in the circularity that associates and is understood as a *bucle* (loop).

The change in the logic of the continuum of causes and effects, including, for example, the principle of uncertainty, will cause a paradigm shift at the beginning of the 19th Century called "delocalization of the epistemology" promoting new questions and new interpretative locations because it covers other fields or subjects/ disciplines and inspired authors such as Lupasco, who published works introducing the subject in *Trois Matières* (1963) and after developing in *L'Homme et ses Trois Ethiques* (1986) about an ethics of the psyque (Weil, 2000).

It is not surprising that the objectivity and stability of the method of physics that used to prevail in the paradigm was naturalized and universalized, but now under a new logic that will allow for the development of the theories of system, of complexity and, consequently, of transdisciplinary, have their frames of reference in the paradigmatic and epistemological movement made by physics. In this respect, to understand that reality is composed of several levels, to ignore the limits of the subjects/disciplines and to approach phenomena taking into account the relations that they propose, will compose the premises that this new paradigm contemplates. This paradigm had an impact in different fields such as social sciences, when considering concepts like educational systems, language emergency and complex communicative interaction (Flórez et al., 2017).

The fact is that studies of these levels of reality have been shaped in the history of science, from biased research to an integrative form of investigation. This logic of integrated science will enable the constitution of an environmental science. In effect, disciplinarity, multi-disciplinarity, interdisciplinarity and transdisciplinary will be the four ways in which knowledge is produced. The fixation of one of these ways will determine scientific practice and, consequently, the degree of knowledge of the levels of reality and the articulation of these levels will enable the systemic understanding of the human.

To distinguish the previously described, it is necessary to specify the following properties: according to the history of science and epistemological debates, macro-fields have been created with a general taxonomy that identifies and organizes sciences and subjects/disciplines according to their objects or facts of study. As a consequence, science has been characterized in four major groups: formal sciences, natural sciences, human sciences and social sciences. Each one of these branches will be integrated by subjects/disciplines and from these, from the specificities of the "objects" of study, associated taxonomies are displayed.

Taking into account the above, while disciplinary refers to the organization of scientific knowledge from fields or specialized areas of knowledge, multidisciplinary (parallelism) is the study of a disciplinary object by several subjects/disciplines at once. And while the interdisciplinary (perspectivism/convergence) is the *transfer of methods* (Basarap, 1995) of the subjects/disciplines that configures the same field; transdisciplinary "is interested in the dynamics of the action of several levels of reality in face of the understanding of the present world" (Osorio, 2011, p. 285), what can be called holism.

This transdisciplinary approach understands reality as multidimensional, structured in multiple levels, and will replace the one-dimensional reality discourse. Pombo (2013) suggests that the distinction of science between pluri-, inter and trans-disciplinary should be thought of as a *continuum* that goes from coordination to combination and from combination to fusion.

"If we join this continuity in the form of a growing intensity, we will have anything of this gender: from pluridisciplinary parallelism to perspectivism and interdisciplinary convergence and, from this, to holism and transdisciplinary unification" (p. 25).

This logic of the *continuum* established by Pombo allows us to understand how the speeches and environmental studies were constituted; that is, that the constitution of what is going to be called environmental science, will be forged through a theoretical statute that evolves from the disciplinary to the interdisciplinary and then to the transdisciplinary. But, beyond these new types of disciplinary configurations, Pombo (op. cit.) states that it is still possible to point out the existence of new practices of interdisciplinary crossing (crossing, convergence, decentration). The environmental issue would lead to decentralization practices, so it is irreducible to a disciplinary field.

To understand the complexity of the society in relation to the fields of scientific knowledge from the human sciences, especially since 1980, various authors began to build bridges between natural sciences, life sciences and social sciences. Dorado et al. (1997) points out that the challenges of interdisciplinarity are posed in the following terms: the first challenge points to interdisciplinarity between natural sciences, life sciences and sciences of man. The second, the relationship between life sciences and human sciences. And, finally, the third challenge consists in facing the interdisciplinarity between the different sciences of man or social sciences that, according to the author, include: anthropology, history, ethnology, sociology, politics, etc.

The Theory of Systems will be the bridge between natural and social sciences, since it is integrated into a transdisciplinary methodology that will recognize the social field. In this direction, environmental science will be possible because it goes from a simplistic view where living beings are studied only from a physical-biological perspective to a vision that takes into account the social context. The complex approach will encompass both physical and biological contexts, as well as social context.

It should be noted that the concept of environment formulated from an ecological point of view refers to its origins to the physical and chemical conditions of a place, into which living beings are inserted and form ecosystems. In view of this, the conjunction between the triad: abiotic, biotic and anthropic, comes to configure an interpretive framework of the environment from its systemic conjunction, in a network. This conjunction will allow us to talk about transdisciplinary as a strategy to face the complexity of this system, allowing us to go *beyond* the disciplinary field.

Understanding the environment from a specific field of knowledge creates vices and voids due to the fragmentation of interdependent networks of the environment; that is, when carrying out a study under a disciplinary object, and starting from that, problematize it under the conceptual domain of the subject/discipline, and that breaks with the logic of the system and the complexity on which it is sustained.

According to Giannuzzo (2010), the environment is conceptualized in the following ways: in law it is recognized as a product and in economy as an environment; for chemistry and engineering, it refers to the abiotic components of ecosystems; in biology and ecology, as biotics and abiotic elements; in social sciences, to indicate external relations to a phenomenon under study. In this regard, the author states that there is an absence of a consensual epistemological foundation on the environment, originating in inaccuracies in its theoretical foundation.

From an environmental epistemology perspective, the environment is conceptualized as the field of relations between nature and culture, "what is redefined is the dynamic link between both observable affections" (Giannuzzo, 2010, p. 142). In this regard, Torres (2016) considers environmental epistemology as follows:

> It recognizes the different dimensions and interpretations and chronicles and its form of return (Morin 2003), and it has prevailed the idea of understanding it in its simplest form – that is, equal to ecology –, reducing it to the natural environment, above other dimensions – in its different metamorphoses and mutations – (p. 45).

2.2.1 From the Speech of Interdisciplinarity to a Practice for Development

After the main problems in the aftermath of the Second World War (1945), the world needed cooperation in various fields: economic, political, scientific and cultural in order to organize and overcome the fractures left by irrationality—rationalized by the fetishism of the *ethos* and its social practices of that period. The compartmentalization of the subjects/disciplines would not be possible due to such fractures, for which UNESCO, at the end of the 1960 promoted research from an interdisciplinary perspective. The Acts of the 15th Session of the 1968 General Conference in Paris demonstrate this.

These documents demonstrate the interdisciplinarity speech anchored within the objectives for development. With regard to scientific policy, scientific information and fundamental research, UNESCO (1969) seeks to "promote a coherent national scientific policy that will increase the scientific and technological potential of the country and guide its scientific activities towards the satisfaction of the needs of development" (p. 34).

To UNESCO, new centres responsible for cataloguing, research and training must be created from an applied interdisciplinary perspective. In this direction, the environmental sciences and natural resources, life sciences, earth sciences, social sciences, human sciences and culture, as they were denominated at that time, should allow for the strengthening of the states. In view of this, the UN establishes or "invites its member states" to follow the directions for each of the referred fields.

For example, in the environmental sciences and research on natural resources, they are asked, in particular, and following scientific bases, to carry out a qualitative and quantitative study of their natural, terrestrial, non-oceanic aquatic resources, to encourage the acquisition of knowledge in all branches related to the possible use of plant and animal resources, allowing the enactment of legal provisions for the use of these resources, but without losing sight of its conservation and rational use; also, the study of the physical environment of the earth and especially of the geophysical phenomena originated by solar activity, the upper geological layers of the earth; the causes, the mechanism and the prediction of earthquakes and phenomena related to volcanology, among others.

For this, experimental stations and a network of appropriate institutions for the promotion of interdisciplinary research and the formation of human capital in geology, geophysics, geochemistry, edaphology, geomorphology, ecology, hydrology and oceanography, among others, should be created (UNESCO, 1969).

In relation to social sciences, these should promote interdisciplinary studies on the socio-cultural and economic factors and effects of development, the economic and socio-cultural examination of the problems of education, science and technology, ensure effective implementation of the rights proclaimed in the Universal Declaration of Human Rights and facilitate the widest dissemination of the conclusions of science on the unity of the human species (UNESCO, 1969).

From the various points that are evident in the aforementioned declaration, there is one that is particularly interesting: the conception of the *medium*. By having these various meanings because it is subjected to different explanations, sometimes to a biological conception and other times to a social, psychological, spatial, aesthetic conception, UNESCO indicates that its content must be specified. Articulating with interdisciplinarity speech, this conceptualization must be approached from several disciplines integrating, therefore, the social and human sciences and the exact and natural sciences, emerging integration under the slogan: Man, and his medium/environment: bases of a better life.[1]

The concept of environment (*in lato* sensu) is not new, based on the foundation of evolutionary biology of Darwin in the publication The Origin of Species in 1859 and the movement that followed: Darwinism, "revitalized" in order to support geography as a subject/discipline, especially to human geography, when trying to solve the crossroads situation in which such subject/discipline found itself. One fact called attention to the period of 1940–1960 in relation to disciplinary geography: the existence of a certain gap within geographical theory about the events of this subject/discipline in the war-torn area in the post-war period. What has been evidenced is a loss of protagonist in political geography that reappeared in the 1980s with the disassociation of geopolitics. In this regard, the environment conceptualization developed at the end of the 19th Century, both in the natural sciences and in the nascent social sciences, and reappeared in the 1940s and 1960s of the 20th Century embedded with greater property through what was called anthropogeography, in school geography and within the subject/discipline geopolitics (Taborda, 2015, p. 159).

Guided by evolutionist philosophy, starting with Darwinism, it will be intuited in geographical studies that the environment is no longer a natural spot in the sense of biology, although it is assumed, under a mutation: geographic environment and under the theorization of the existence of a "social organicism". To that extent, the birth of human geography was developed by accepting the category of environment from evolutionary theories. The emergence of the environment within geography is

[1] The joint group of the commission noted that recommendations of the Biosphere Conference provided UNESCO with a precise framework of action, focused namely on Exact and Natural Sciences, but for which it could be convenient the participation of specialists in social sciences and education. These recommendations called for UNESCO to establish, in 1969 and 1970, a long-term plan for a scientific, technical and educational program for the rational use and preservation of biosphere resources, which would be submitted to examination in the General Conference at its 16th meeting, along with proposals related to the coordination of the program in international jurisdiction. The joint group considered that the international program resulting from the Biosphere Conference could constitute the core of a set of activities related to the general theme "Man and its Medium" (UNESCO, 1969).

evidenced by three facts: the value of the experimental method to define sciences, the publication of Origin of Species by Darwin in 1859 and the implementation of [the concept of] social organicism.

Although it is not possible to equate the concept of environment in the post-war period, in the middle of the 1940s, with biology or geography (region and territory perspectives) (Taborda, 2015), it was given a possibility, through environmental theories and the influence of nature, to classify environment whether as a determinant or an enabler and under an adaptation perspective, in the social constitution itself; that is, there is an integration of humankind in the milieu, thus generating a symbiosis between the anthropogenic environment *vis-à-vis* the natural environment, and in this relation, unequal or not, environmental discourse starts to shape.

Returning to the interdisciplinary topic and taking into account the fact that the harmony between humankind and the environment is seriously threatened by certain consequences of the evolution of techniques. Especially because of industrialization and other activities, it is necessary to project possible effects in science (UNESCO, 1969). The relation between interdisciplinary science and development is again evident, and not as a condition for development, but to evaluate possible antagonist effects:

"Notion and a general sense of the order of priority set by the General Conference, namely, in 1960, education, and, in 1964, natural, exact sciences and technology, as this order of priority corresponds to the fundamental requirements, must be maintained. However, one could not imagine the development without the progress of the sciences that deal with man and culture (UNESCO, 1969, p. 86).

The 1969 Acts from UN enabled the interpretation of what had happened. For the constitution of the interdisciplinarity speech, while advocating the "transfer of methods" in the subjects/disciplines, there was required compartmentalization or a disciplinary establishment, taking into account the approach used to understand and compact the "object" of your study. It was necessary to segment the disciplinary fields so they could dialogue. But geography was (is) interdisciplinary.

With a definite purpose of articulating the subjects/disciplines of a field of study that would respond to the prevailing needs of development, the interdisciplinarity speech was promoted. It seems that, for the achievement of the objective, it became necessary to make a statute that disciplined the sciences; for this reason, large groups were defined and classified: fundamental sciences, life sciences, earth sciences, social sciences, human sciences and culture. However, the interdisciplinary vision encouraged by UNESCO would be classified as above, for which specialties were developed, such as, for example, geochemistry, ethology, among others, within earth sciences.

Both earth sciences and life sciences are responsible for promoting research on the disposition of natural resources and the internal dynamics of the earth: volcanology, seismology. In the case of social sciences, they should promote studies on the social and economic aspect of the role of education, science, technology, culture and information and development, the place that corresponds to the social sciences in national scientific policy, the development of education and demographic evolution, themes related to peace and how Man can contribute to transform its medium.

The way that natural and exact sciences are articulated with social sciences is diffuse. That is the reason why the 1968 Acts of Paris identified that the discourse on the relation between man and medium lies in the latter. It is not fortuitous that the evaluative role of development policies lies precisely in social sciences. Upon promotion, interdisciplinary research inevitably results in a multidisciplinary approach[2]; that is, involving the knowledge of a fact or phenomenon of study from various subjects/disciplines. In this regard, social sciences were called upon to think over the man-medium relation, which does not happen with exact and natural sciences. One would think then that interdisciplinary speech would require defining and delimiting subjects/disciplines, hierarchizing them.

To reduce everything into physics, biological and psychic was the best that was created from the beginning of the 20th Century. That was how scientistic ideology arose. In the course of the 20th Century, the complexity is nourished by the explosion of disciplinary research and, at the same time, its complexity would determine the acceleration of the multiplication of subjects/disciplines (Basarap, 1995).

It was announced that the world panorama, after the crisis of the wars, required the integration of the subjects/disciplines. For this reason, the UN articulated an investigative exercise with interdisciplinarity discourse. This articulation required disciplining science and then articulating it. Also, it was stated that the emergence of theories associated with complexity opens into transdisciplinary speech. This disciplinary articulation and the environmental crises of that time created the panorama for the structuring of environmental science.

2.3 Emergency of Resilience: Changes of Speech

In this topic, will be analysed how the speech of the environment anchored to development will create a network of meanings that will be established in the political consciousness of the authorities and the academic community, through the aims of the summits and conferences made by the UN.

No doubt, national and international agendas have focused on environmental problems, and developing speeches promote the realization above all authorities, of political actions to mitigate the indiscriminate consumption of the "natural capital" of the earth, the promotion of practices that do not threaten the future stability of it and the reduction of climate change.

---

[2] "The general principles of the exact and natural sciences in their application to development, noting with satisfaction the increasing cooperation between the social sciences and the exact and natural sciences, insists that such cooperation should be carried out in a multidisciplinary spirit, and recommends that the science sector participates actively in the future program regarding man and his medium, interpreting these concepts in their broadest sense". It is added: "Considering that it is essential to conserve its priority [education and scientific and technological training] to the development and modernization of science and technology education and to resort to a flexible and innovative multidisciplinary method". (UNESCO, 1968, p. 99-100).

2.3.1 From Ecological to Social (Communitarian) Resilience: Tensions for a Capacity/adaptation Relation

After the 1992 Rio Declaration, a euphoric and optimistic atmosphere was created aiming to confront the onslaught of capitalism, to promote the emergence of an approach within sustainable development, to create correspondences and equivalences between economic and social needs, cultural diversity combined with a healthy environment. It should be noted that this summit had promising precedents indicating that the discourse and actions had already been gestated, as, for example, in the Stockholm Declaration on the Human Environment (1972), the Climate Conference in Geneva (1979), the Kyoto Protocol (May 1992), among others.

One of the principles (out of seven) that resulted from the Rio-1992 Summit is that "States must cooperate in a spirit of global solidarity to *conserve, protect and restore* the health and integrity of the Earth's ecosystem" (p. 5). In view of this, sustainable development became the new planetary paradigm that, undoubtedly, impacted countries with a weak institutional and juridical environmental framework. It is possible to verify that global goals for the environment turned out to be a reference regarding its *protection* and also climate under the control of carbon emissions, the full management of water resources, marine and coastal affairs, forests, and others. But in the face of growing global concern about the serious consequences caused by natural disasters and catastrophes, such as in the last two decades: the Tsunami in the Indian Ocean (2004), Hurricane Katrina (2005), the Kashmir (2005) and Haiti earthquakes (2010), The 2011 earthquake and tsunami Tohoku-Fukushima, and Amazon and Australia fires in 2019. Imbalances and anomalies occurred, especially in the poorest communities, the issue of risk positions again a speech, but now under the relationship vulnerability/restoration as its constituent.

That is why, in 2005, the world conference on disaster reduction took place in Hyogo (Japan), which resulted in the creation of a framework for action (2005–2015), for states and communities to be *warned, prepared and respond* effectively to a disaster or vulnerable event and, also, should they occur, activate the mechanisms that enable their recovery and rehabilitation, that is, to be resilient.

The argumentative vertebra of the speech will focus, then, on the resilience both at the state as well as the community level, assuming risk and vulnerability under a proactive position. In response to this, the UN document welcomed the conceptualization made by the International Strategy for Disaster Reduction—ISDR, understanding resilience as "the response capacity of a system, community or society potentially exposed to threats to adapt, resisting or changing [...]" (p. 9).

This definition, according to Trujillo (2011), is problematic and restrictive, considering that the condition of possibility of a community is equated with the aptitude and virtue that, translated into an ethical–political level, could lead to inequitable actions. On the one hand, because it would be restricting the state action to teaching resilience, and, on the other hand, the level of social justice would come only from a communitarian determination that would constitute poverty and oppression as an internal dimension and to which they have not been shown how to be resilient.

That is why various authors such as Rutter (1991), Luthar and Cushing (1999), Kaplan (1999) define resilience as a process (dynamic between emotional, cognitive and sociocultural mechanisms) diachronic and synchronic, at the same time, giving space to the historicization of the subject. According to Trujillo (2011), this procedural perspective of resilience:

"it allows to overcome essentialist, timeless understandings, opening a place to the future of certain approaches, giving space to complex, historical, dynamic, existentialist understandings, including ontogenetic development, without having to leave aside structuralist considerations around the elements, qualities, or relationships that shape resilience, which can be subsumed and articulated with each other" (p. 16).

This conceptual tension reveals, in a very evident way, the study carried out by the Colombian research group called Resilio. When analysing 74 academic documents from 14 universities in the country, it was found that, while 28 of them conceptualize resilience as capacity, 21 of them refer to it as a process (Trujillo, 2011). This differentiation is showing that its conceptual construction is still in process, but even so, the aforementioned group has defined it as a capacity, as has already been demonstrated.

The concept of resilience as a combination of factors that allow a human being to face and overcome the problems and adversities of life, began to develop from a psychiatric and later in a psychological perspective, mainly in Europe and in the United States. These fields constituted the conceptual bases of resilience, when demonstrating that people, despite living in traumatic, dysfunctional or disadvantaged contexts, were able to develop psychologically in a stable manner. That was the outcome of the research carried out by Rutter.

In the early 70's, Rutter put the concept of resilience in the field of psychology, counteracting the concept of invulnerability that was at its peak. According to Rutter (1991), the emergence of the concept of resilience versus the concept of invulnerability occurs due to "the data we had over individual differences between risk situations, studies on temperament and differences in facing life situations" (Puig & Rubio, 2013, p. 34).

Holling (1973), in his ecological studies, Resilience and Stability of Ecological Systems, introduced the study of resilience. In that study, resilience evidences the ability of biotic communities and ecosystems to recover from disturbances and impacts while maintaining certain characteristics of structure, stability and functionality. The author proposes an adaptive cycle in four phases: growth, conservation, liberation or creative destruction and reorganization, allowing him to understand how systems can adapt and thrive, while adapting to change. From these "indicators" or phases will be determined whether ecological systems have high or low resilience (Holling, 2001; Holling & Gunderson, 2002 Eschenhagen, 2007).

In this respect, Castilla and Velázquez establish that these four phases also allow us to understand different systems such as ecosystems, social systems, institutional systems and socio-ecological systems. The meaning of natural disaster, whose impacts make ecological systems vulnerable and, therefore, the society that territorializes them in their political, economic and social interaction, will begin to resignify itself and mutate from vulnerability as weakness, to a relational and social conceptualization of the disaster, a meaning that will be intertwined with that of resilience under a process of coping (absorption in ecology) adaptation and transformation.

According to the above mentioned, in the studies on resilience one can distinguish two approaches or traditions: the geneticist-individualist and the interactional—ecological. While the first, from the beginning of the 70's, describes ways of dealing with psychological tensions, especially in children and young people submitted to traumatic events, in a positive way: positive adaptation to adversity; the second refers to the capacity of ecosystems, under a dynamic process in which environment and human beings converge, to *recover and return* to catastrophic events.

In this second approach, especially since the 1990s, resilience is expanded "by the notion of process, which involves the dynamic interaction between risk factors and resilience, and the search for models to effectively promote Resilience at the level of social programs" (García & Domínguez, 2013, p. 67). As evidence, the focus of interest unfolds from two fields, one referred from personal qualities and another with a greater interest in external factors, but assuming the psychological level within the resilience assumed as a process; is affective and cognitive.

It has been briefly described that the environmental theme during the last two decades has been supporting the international agenda, which has enabled many countries to register and consequently take political action under the discourse of sustainable development. But, given the impossibility of effectively reducing vulnerability, it is argued that societies and communities must be resilient.

To that extent, speeches on risk management are understood from a resilience perspective. Starting from the conceptualization point of view, it has also been said that conceptualizing resilience as capacity will result in ethical–political tensions, because it demobilizes communities. Therefore, it is necessary to determine at which moment the discourse of resilience is articulated with the discourse of sustainable development, or, on the contrary, whether the latter has always been contained in the discourse of resilience or whether it is mutating to the discourse of resilience.

The thesis that arises is that, although there is no concept of unified resilience, within the epistemological field in which it is positioned, the truth is that its meaning is bounded with other concepts that constitute it and that, in turn, position it between risk factors and protective factors at the level of an individual, a community or a society, these being those of confrontation, transformation (regenerative) and adaptation to adversity. But as the category of resilience emerges from one field to another, tensions arise, allowing us to see that by transcending this from psychology to ecology and then to social, and in the absence of ecological and social justice, it is assumed there is an adaptation in the social field, bringing to life a know-power speech.

To this extent, it would be worth asking if there is a close relationship between the discourse of sustainable development and the discourse of resilience or, on the contrary, if the unsustainability of disturbed ecosystems (due to the *"natural capital"*) emerges from the frontiers of sustainable development, but within the concepts of confrontation, transformation and adaptation.

2.4 Sustainable Development and the Discourse of Social-Communitarian Resilience: Frontiers or Changes?

As a result of the social instability deriving from the "world wars" and with the goal of guaranteeing peace, the UN became an official organ on October 24, 1945. In order to lay foundations for international stability, it was necessary to articulate the discourse of peace with a discourse on development; that is, the possibility to rebuild and reinvent in the face of the visible and disastrous consequences of a self-referential system from 1942. Since then, the UN has become a "governing body" responsible to establish policies or guidelines in order to promote and protect the human rights declared in 1948, as well as in the International Covenant on Civil and Political Rights[3] and in the International Covenant on Economic, Social and Cultural Rights,[4] both from 1976.

Along with the foundation of the UN, scientific research has begun to account for a problem never before addressed (or one that had not been given due attention): that environment was vulnerable to human intervention.[5] Studies back then demonstrated this: in the 50 s, photochemical smog pollution was discovered, a pioneering article on the "greenhouse effect" was published, systematic measurement of atmospheric

[3] Adopted and opened for signature, ratification and adhesion by the General Assembly and its Resolution n. 2200 A (XXI), from December 16, 1966. Entry into force: March 23, 1976, according to article 49.

[4] Adopted and opened for signature, ratification and adhesion by the General Assembly and its Resolution n. 2200 A (XXI), from December 16, 1966. Entry into force: January 3, 1976, according to article 27.

[5] The speech of the Green Revolution was initiated by the Rockefeller Foundation and established with the Mexican Program of Agriculture in the 40's (1943) with the creation of the International Center for the Improvement of the Corn and Wheat (CIMMYT) in Mexico. Years later, similar projects reached almost all Latin America through the United States Department of Agriculture (USDA). Other private foundations "sheltered" other territorialities such as Ford Foundation in India, International Institute on Rice Research (IRRI), created with Rockefeller's support, in the Philippines (Ceccon 2008). Regarding to the Green Revolution, International Food Policy Institute (IFPRI 2002) refers to it as "the development of high yielding varieties (VAR), obtained to respond to inorganic fertilizers and other chemical products […] contributing to the existence of surplus of agricultural products and eliminating hunger".

The speech of the Green Revolution will be articulated in 1966 at the World Conference on Agricultural Reform, along with UN and FAO. The 1968 annual publication of the FAO states that "food and studied the increase of agricultural productivity through "technological improvements" as a way to liberate lands that would be used to feed people. The fact is that the Green Revolution required the decrease on the use of agrotoxics and the "seemingly effective" elimination of pests. Ceccon (2008) shows how, between 1963 and 1973, in Brazil, in the face of an increase in agriculture, there was an increase of pests and, therefore, in the use of agrotoxics. That is why, in 1962, Rachel Carson presented alarming data over food contamination by pesticides and consequences due to the use of DDT (Dichloro Diphenyl Trichloroethane) in the United States. The genetic uniformity and the promotion of the use of certain technologies would wreak havoc throughout the world: loss of genetic variety and organization of monopolies. In view of this situation, the storage of germoplasm would be suitable for new extractivism. (Borlaug, 1972).

CO_2 began, and in the 1960s, the Silent Spring (Vázquez et al., 2014) that would inspire the environmental mobilization in the United States (Murphy, 2005). From now on, the UN will organize conferences and summits with the aim of laying the foundations for development that is articulated with the "rational use" of natural resources, while evaluating the problems of man with his environment.

Tracing the resilience in the summits and conferences carried out by the UN the following is found: neither at the UN Conference on the human environment in Stockholm (1972) nor in the Brundtland report "Our Common Future" (1989) or years later in the Rio Declaration (1992) or in Johannesburg (2002), the resilience speech is present; the main issue addressed is development and from the Brundtland Report on sustainable development.

In the Stockholm Conference (1972), for example, was established the need to address the problem of environmental deterioration, but in a perspective of (economic) development. Sustainability is not the focus, only under the commitment to preserve, conserve and improve the environment, through planning.

However, organizations that organize their resources rationally should be created. As can be seen in Fig. 2.1,[6] the categories that are related to the development discourse are determined by the principle and the need to generate economic conditions that makes it possible, so it is necessary to produce a scientific plan for the interdisciplinarity of science, as shown in the topic 2.2 above.

With the Brundtland Report, edited by UN (1987), was created the expression "sustainable development", defined as "the development that meets the needs of the present without compromising the ability of future generations to meet their own needs". Such Report establishes the basic premises for the sustainable management of the earth's resources: to give priority to basic needs (food, clothing, housing and work) of the poor in the world and to emphasize that the limits of development are not absolute but conditioned by: (i) the state of technology at the time, (ii) the organization of the society, (iii) the impact on the environment and (iv) the ability of the biosphere to absorb the impacts of human activities. In, the associations of the words with the development category are identified, by its proximity to the main core, with the sustainable category.

Twenty years after Stockholm and three years after the Brundtland Report, at the UN Conference on Environment and Development, held in Rio de Janeiro in 1992, it was agreed that the protection of the environment social development and economic development were fundamental to achieving *sustainable development.* In order to achieve these goals, "Agenda 21" was approved.

This agenda enables an emerging speech on sustainable development, since it creates an international agenda that raises fundamental discussions in social and

[6] The graphics presented were made through the T-LAB computer program. The software is composed of a set of linguistic, statistics and graphics tools designed for the analysis of texts. Methodically, it identifies with a speech analysis and methodologically it is developed in the following way: the cited reports of the conferences are archived in the system, the relations of co-occurrence between the key words are explored and measured, the lexical units and semantic maps of word associations and relationship networks are then generated.

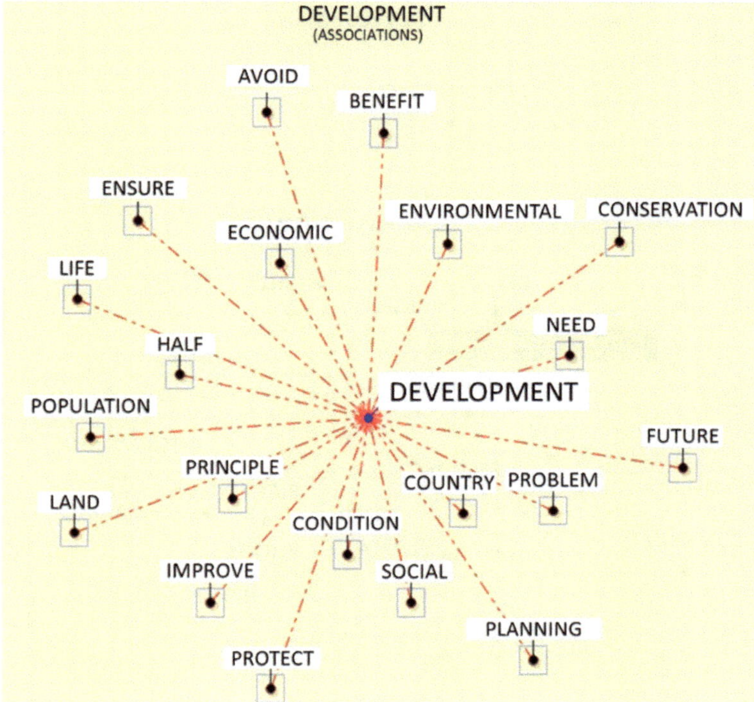

Fig. 2.1 Semantic network of the 1972 UN Stockholm Conference. *Source* Own T-Lab

economic dimensions, in conservation and management of resources for development, in strengthening the role of the main groups and the means for their execution (IFPRI, 2002).

That is why the fight against poverty, the promotion of human health, the integration of the environment and the development of the decision- making procedure, the protection of the atmosphere, the fight against deforestation, science, among others, comes to nurture and position the discourse in a way that rules cannot ignore, that is, those are subjects that politician must address in their territories,

Allowing them to overcome inequality gaps, since all environmental imbalances generate unequal mechanisms that makes a *good living* impossible in the communities (Fig. 2.2).

Twenty years after Rio-1992 a new summit takes place in the same location. The agenda for sustainable development, through the promotion of a sustainable economic, social and environmental future for our planet and also for present and future generations, still stands, as well as the definition of sustainable development as presented in the Brundtland Report. In Rio-20, politicians renewed the agreed compromise made in Rio-1992 and, in addition, agreed to comply with those purposes, through concrete and urgent measures, since they recognized that

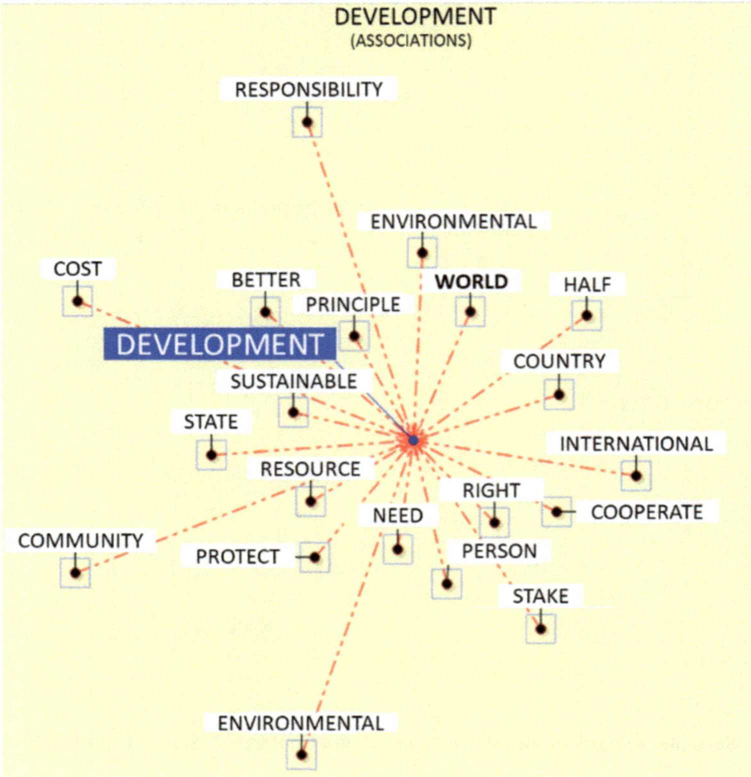

Fig. 2.2 Semantic network of the 1992 UN Rio de Janeiro Conference. Graphics designed by the authors powered by T-LAB

since 1992 progress has been insufficient. The eradication of poverty was one of its determinants.

Twenty years later, leaders still talk about present-future (space-temporality); but there was no talk of resilience, as shown in the Fig. 2.3, nor about the green economy. Regarding resilience, twelve references are made in the document, in which nine are under different items.

It is not fortuitous that the protagonism of the placement within the report is constituent of resilience, a relation that will be analysed when communitarian resilience is addressed, where it tends to transcend this concept, even at a theoretical level. Such a concept, observing the discussions and fields in which they are being articulated, defining and redefining them, is redimensioning a categorial level.

This Rio-20 report will embrace the guidelines established ten years earlier in the Hyogo Framework (2005), and this framework is articulated with the Strategy of Yokohama (1994), a strategy in which resilience is not nominal, but if it is the disaster under the triad *prevention-reduction-reparation.* It is necessary to highlight that,

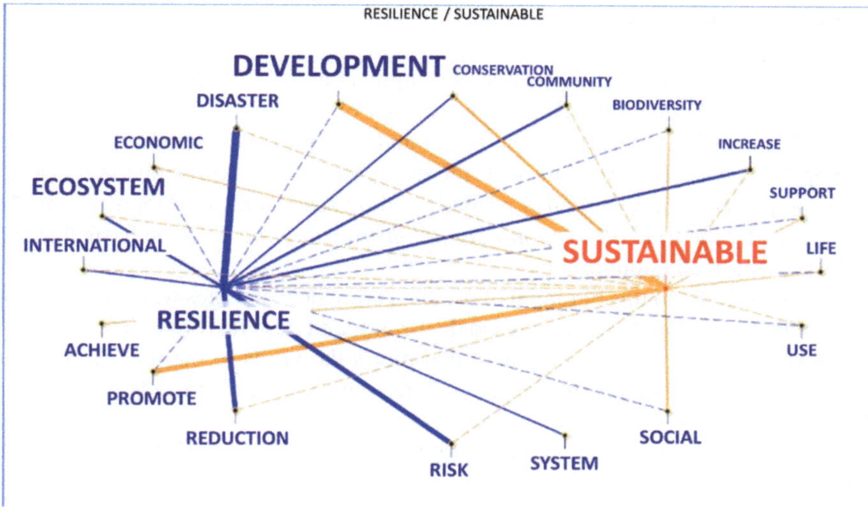

Fig. 2.3 Level of relationship between resilience and sustainable development in the report of Rio + 20 Conference in 2012. Graphics designed by the authors powered by T-LAB

within the Hyogo Framework of Action, resilience is articulated with the objectives of sustainable development, but with greater priority in the risk of disasters.

In this regard, the Hyogo Report recognizes that "human societies must live with the risk posed by the dangers of nature. However, we are not at all incapable of preventing and mitigating the effects of disasters. We can and must alleviate the suffering that those dangers entail by reducing the vulnerability of societies" (UNISDR, 2015). A culture of disaster prevention and resilience must be promoted at all levels, from the individual to an international level, as well as strategies prior to disasters.

In this context, the Resolution approved by the General Assembly on September 25, 2015, where it defines the 2030 Agenda for Sustainable Development, the Sendai Framework for Disaster Risk Reduction 2015–2030 and the Framework Convention on Climate Change[7] held in Paris,[8] and embraces the discourse of resilience. In the first document, resilience is emphasized in the habitats, the infrastructure, the communities that host refugees, in human settlements, the poor,[9] in agricultural practices in the face of climate and disaster risks and marine ecosystems. In the

[7] The UN Framework Convention on Climate Change (UNFCCC) is part of the three treaties enacted at the 1992 Rio Summit (the other two are: Convention of Biological Diversity and Convention to Combat Desertification); coming into force in 1994. Among its main objectives is to stabilize the concentrations of greenhouse gases (GHGs) in the atmosphere in order to prevent risks in the climate system. Up to date, 23 COPs have been made.

[8] Conference of the Parties. 21st session, Paris, November 30. FCCC/CP/2015/L.9.

[9] "Until 2030, it is important to foster resilience from the poor and people in situation of vulnerability and reduce their exposure and vulnerability to extreme phenomena related to climate and other economic, social and environmental disruptions and disasters".

second document, in addition to those above stated, it applies to sanitary, cultural and educational.

Seeking a link between sustainable development and resilience, it is evident that this link has no boundaries. But while the idea of resilience is still in an embryonic stage, it embraces however the initial and operational concept of sustainable development; what is happening with the emerging articulation is that the triad to *conserve, protect and re-establish* promulgated by sustainable development should be replaced by the triad *absorption-adaptation-transformation* on the basis that the risk is present in all communities.

How to understand this binomial of triads? The answer emerges when the concept of sustainable development presented in the Brundtland Report is brought up: sustainable development is "the development that meets the needs of the present without compromising the ability of future generations to meet their own needs" (UNISDR, 2015).

Studies of the impacts of global warming (WHO, 2016, NASA, 2017, IPCC, 2014) show that the ecological footprint, or the degree of impact of human beings on the planet, has compromised future capacity in the present. Convincing evidence confirms, albeit conservatively, the reports of the Intergovernmental Panel on Climate Change of 2007 and 2014 (Field & Barros, 2014, Stocker, 2014). These reports indicate that the discourse of sustainable development, although it has promoted political practices and, consequently, institutional improvements, has not redefined the economic-political system on which the degradation of nature is based.

To that extent, the difference between resilience and adaptation to climate change is becoming increasingly diffuse, as is the notion of vulnerability and risk reduction. While they have this diffuse relation, the idea of sustainable development seems to be changing towards a discourse that is aligned with the speech of resilience.

2.5 Resilience and Biocultural Heritage: Sketches from Political Ecology

As said before, the speech of resilience in the social sciences raises mistrust, on the one hand, because the relationship between vulnerability-risk is countered with that of resilience as a way of rebalancing failure, and this, on the other hand, raises, at the state level, the provision of mechanisms to monitor communities, which, are being asked to be resilient to poverty and disasters, not before being delivered the discourse on socio-environmental justice and good living as political practice and as its constituent; that is, not distanced from that of resilience. Communities manage their own difficulties themselves.

Resilience in the communitarian field, instead of being a system of confrontation-adaptation-transformation, must be understood and developed as a system of organizational irruption; that is, the ethical–political commitment of the communities. To that extent, the passage from the ecological and psychological resilience discourse to one of social resilience could not be extrapolated without first challenging the political ecology of extractivism, in order to analyse the processes of meaning, valuation and appropriation of the nature in line with *the ecological debt*. Leff (2003) analyses:

"The political ecology is located in the boundaries of the environment that can be recorded and internalized in the paradigmatic space of the economy, the valorisation of natural resources and environmental services. Political ecology establishes itself in this space that is the conflict for reappropriation of nature and culture, where nature and culture resist to the homologation of incommensurable values and processes (symbolic, ecological, epistemological, political) and be absorbed in terms of market values".(p. 21).[10]

Political ecology is the gamble to the reappropriation of nature with environmental justice and the perspective of good living not only at the practical level, but also at the discursive level. Considering the politicization of resilience and communitarian resilience as a socio-environmental discourse, it deserves its deconstruction, since it is supporting and at the same time legitimizing actions and processes that can generate ethical–political contradictions, inscribing themselves in a logic of know-power: control device. To that extent, risk and vulnerability might reduce risk and communitarian resilience shall be the new form of association with ecosystems, but under a horizontal and asymmetric relationship. And it is asymmetric because the world is now encompassed in the cosmos. It is this fair asymmetry that will allow us to rehabilitate the vital systems of humanity, the biocultural heritage.

According to the above, it is necessary that before the conceptualization that involves socio-community resilience as capacity-adaptation, there is a transformation, managed from a political ecology perspective and anchored in a biocultural heritage, every time this is conceived as an interdependent system that is emphasized in a body as a fact of the territory and in the territory as a collective history. Castillo and Velázquez (2015) state that while "adaptation can be understood as alterations of social relations and alteration of the society within the ecological sphere, but maintaining its structure, transformation may imply alterations of social relations, but always accompanied by changes in social structures" (p. 21).[11]

To this extent, the dynamization of biocultural heritage does not imply subjecting it to the system that destroys it. Rather, by situating the Otherness of the people of the world as an organizational breakthrough in the face of modern economies, it implies criticizing the system that destroys or subsumes it, while affirming and promoting a good life for communities through its biocultural heritage. Insofar as

[10] In the original: "*La ecología política se localiza en los linderos del ambiente que puede ser recodificado e internalizado en el espacio paradigmático de la economía, de la valorización de los recursos naturales y los servicios ambientales. La ecología política se establece en ese espacio que es el del conflicto por la reapropiación de la naturaleza y de la cultura, allí donde la naturaleza y la cultura resisten a la homologación de valores y procesos (simbólicos, ecológicos, epistemológicos, políticos) inconmensurables y a ser absorbidos en términos de valores de mercado*".

[11] In the original: *"la adaptación se puede entender como las alteraciones de las relaciones sociales y de la sociedad con la esfera ecológica, pero manteniendo su estructura, la transformación puede implicar alteraciones de las relaciones sociales, pero siempre acompañadas de modificaciones en las estructuras sociales."*

food stability is generated in the communities, it allows, from its externality to the system or from its internal transcendentality, to assert itself. That is, for the political ecology to place the ecology in the perspective of an epistemology and a political practice takes account of a double sense in which though liberalism and capitalism can assume biocultural heritage: it denies it as epistemology for a real political praxis and it assumes it under the purpose of valuing what already has a value; that is, to be constituted as *biocultural capital*.

So, the biocultural heritage understood from the political ecology perspective criticizes the two dominant paradigms of modernity: from the economic theory of development and from the biological theory that has normalized socio-environmental dynamics; that is, the ecological distribution before the geopolitics of sustainable development. In this way, the biocultural landscape, conformed by agro-ecological communities and natural ecosystems, the biocultural memory understood as a set of knowledge built from beliefs, knowledge and practices, in integrating with resilience implies an adaptation to the effects of climate change, developmental public policies, an epistemology of predation, but found in the conditions of life that enable it to generate dynamics of social mobilization through the environmental rationality of communities; that is, the establishment of struggles for the reappropriation of their biocultural heritage and the construction of new territories of life. It is the transition from a subaltern culture to an alternative movement before a "low intensity war" and the establishment, before the coloniality of the human being, knowledge and power.

2.6 Conclusion

It has been shown that the constitution of studies on the environment as a relational system between nature and the human being, intertwines with the complexity and transdisciplinary discourse. From various fields of science, the fallibility of classical science, such as biology, ecology, quantum physics, and society's sciences, was found, proposing new epistemologies.

The consequence was the appearance of a new paradigm that, by questioning mechanism and determinism, realized that reality is built out of different levels that interact among themselves. The compression of such levels will require a questioning from disciplinary research, since they focus on a specific level of the system and, furthermore, by the method that defines it, making then impossible for the discipline itself to understand its object of study. Therefore, it was necessary to move from a disciplinary structure to an interdisciplinary or transdisciplinary one. It will be this logic of interrelation between sciences that will allow us to establish an epistemological statute that will make environmental science, environmental studies and the redefinition of the concept of environment possible.

The devastating consequences of the "two world wars" for the appearance of the UN (1945). The founding principle of the UN is to maintain peace among nations and this would require the establishment of policies that would allow, in addition to counteracting the order of the cold war, potentiate the idea of democracy and development. Natural environment and natural resources are the sources that will enable us to sustain the praxis of the idea or the ideology envisaged. There will be then the need for a science and training under an interdisciplinary statute. Gradually, environmental issues began to position itself in the international public agenda, especially in the late 70's. And parallel to this, within psychology, a concept is being theorized that will later transcend environmental studies: resilience.

With the Brundtland Report (1987) emerges the discourse of sustainable development. From this date forward, conferences were held seeking to promote *long-term* goals and specific actions, at national level, to avoid environmental deterioration. But in 2012 at the Rio + 20 Conference, sustainable development was anchored by another discourse: resilience. And resilience has already been conceptualized in the Hyogo Report (2005) as the *adaptability* of a system against disturbances; that is, under risk prevention. It was analysed that conceptualizing communitarian resilience as capacity is inconvenient, because it subsumes the potentiality of the elevation of Other in justice. Instead, it must be assumed as a procedure and, moreover, it must be thought of from a political ecology perspective.

In this way, the complexity of the environment, or rather, the dynamics resulting from the interaction between nature and human being, requires articulation between the knowledge of several disciplinary areas, because it transcends the partial visions of the disciplinary perspective. Even in transdisciplinary studies, it should be noted that "this study has its limitation, as offering itself as a meta-discipline without yet being one, since it must be thought of in terms of its capacity in approaching the environmental problem in a complete manner. Consequently, popular knowledge is left behind or even disregarded, believing that it is able to close the gaps between subjects/disciplines, to the point of having an amplified but 'empty' object" (Torres, 2016, p. 46).[12]

The implications that this has in the ways of proceeding reached, no doubt, the heart of social practices, either by organizations that seek to develop mechanisms, not only to deal with environmental deterioration, but to manage and create communitarian systems, taking into account their vital needs or of ethical–political positions; and the ethical one is approached by a critical vision that, in the case of the environment, is related not to the anomalies of the system that Grosfoguel (2006) calls "imperialist/occidentalocentric/cristianocentric/capitalist/patriarchal/modern/colonial", but in the very constitution of the civilization. That is, they are products of the same system and, therefore, could not be referred to as anomalies.

[12] In the original: "que ésta conlleva limitaciones, al ofrecerse como meta- disciplinaria sin serlo todavía, ya que se piensa tener la capacidad de abordar la problemática ambiental pretendidamente de manera total. En consecuencia, se deja fuera o se desatiende el conocimiento popular; creyendo cerrar las brechas entre las disciplinas, al punto de tener un objeto amplificado pero plano.

References

Armenteras, D., González, T. M., Vergara, L. K., Luque, F. J., Rodríguez, N., & Bonilla, M. A. (2016). Revisión del concepto de ecosistema como "unidad de la naturaleza" 80 años después de su formulación. *Ecosistemas 25*(1): 83–89. https://doi.org/10.7818/ECOS.2016.25-1.12

Basarap, N. (1995). La transdisciplinariedad. Manifiesto (Trad. Norma Núñez-Dentin Gérard Dentin). Monaco: Editions Du Rocher.

Bertalanffy, L.V. (1934). Wandlungen des biologischen Denkens, Neue Jahrbücher für Wissenschaft und Jugendbildung, 10, 339–36.

Borlaug, N. (1972). *Defensa del DDT y otros plaguicidas: Contra el hambre y otras enfermedades del mundo.* UNESCO.

Caro, T., & Alejandra, M. (2015). La geografía escolar en Colombia: travesía hacia su invisibilidad en la segunda mitad del siglo xx. (Tesis doctoral). Universidad Pedagógica Nacional, Bogotá.

Castillo-Villanueva, L., & Velázquez-Torres, D. (2015). Sistemas complejos adaptativos, sistemas socio-ecológicos y resiliencia. *Quivera, 17*(2), 11–32.

Ceccon, E. (2008). La revolución verde: tragedia en dos actos. *Ciencias, 1*(91), 20–29. [En línea] http://www.redalyc.org/pdf/644/64411463004.pdf

Dorado Segura, J. A., Martín Gómez, M. T., Gallero Delgado, J. M., Sánchez Olaya, M. S., Guerrero Gil, J., Tena Ramos, P., & Díaz Acero, P. (1997). Marcha interdisciplinar. http://hdl.handle.net/11162/30799

Eschenhagen, M. (2007). Las cumbres ambientales internacionales y la educación ambiental. *OASIS, 12*, 39–76.

Field, C., & Barros, V. (2014). *Climate change 2014: Impacts, adaptation, and vulnerability: Working Group II contribution to the fifth assessment report of the Intergovernmental Panel on Climate Change.* Cambridge University Press.

Flórez Romero, R., Castro Martínez, J., & Arias Velandia, N. (2017). Comunicación, lenguaje y educación: una mirada desde las teorías de la complejidad. *Revista Folios 30*(2009). https://doi.org/10.17227/01234870.30folios25.38

García-Vesga, M., & Domínguez-de la Ossa, E. (2013). Desarrollo teórico de la Resiliencia y su aplicación en situaciones adversas: Una revisión analítica. *Revista Latinoamericana De Ciencias Sociales, Niñez y Juventud, 11*(1), 63–77.

Gauta, J. (2003). Sistema y pensamiento complejo. En J. Rozo Gauta, Sistémica y pensamiento complejo I. Paradigma, sistemas, complejidad, pp. 49–110. Medellín: Biogenesis.

Giannuzzo, A. N. (2010). Los estudios sobre el ambiente y la ciencia ambiental. *Scientiae Studia, 8*(1), 129–156. https://doi.org/10.1590/S1678-31662010000100006

Grosfoguel, R. (2006). La descolonización de la economía política y los estudios postcoloniales: Transmodernidad, pensamiento fronterizo y colonialidad global. *Tabula Rasa, 4*, 17–46.

Hobsbawn E. J. (1999). Historia del siglo xx CríticaGrijalbo. Buenos AIRES

Holling, C. S., & Gunderson, L. H. (2002). Panarchy: Understanding transformations in human and natural systems. Princeton University Press.

Holling, C. S. (1973). Resilience and stability of ecological systems. *Annual Review of Ecology and Systematics, 4*, 1–23.

Holling, C. S. (2001). Understanding the complexity of economic, ecological, and social systems. *Ecosystems, 4*(5), 390–405.

International Food Policy Institute (IFPRI). (2002). Sustainable options for ending hunger and poverty: Green Revolution Cure of Blessing. http://www.ifpri.org.

Intergovernmental Panel on Climate Change (IPCC). (2014). Climate Change 2014: Synthesis Report. Contribution of Working Groups I, II and III to the Fifth Assessment Report of the Intergovernmental Panel on Climate Change. Geneva: IPCC.

Johannesburg Declaration on Sustainable Development. (2002). In: Report of the World Summit on Sustainable Development, Johannesburg, South Africa, 26 August–4 September 2002. New York: United Nations.

Kaplan, H. (1999). Toward an understanding of resilience: A critical review of definitions and models. In M. Glantz & J. Johnson (Eds.), *Resilience and development: Positive life adaptations* (pp. 17–84). Plenum Publishers.

Armenta, S., Angulo, C., Rocha, W., Barraza, G., Andrade, R., & Gonzalez, J. (2016). Determination and analysis of hotspot areas of deforestation using remote sensing and geo-graphic information system techniques. Case study: State Sinaloa, México. *Open Journal of Forestry, 6*(4), 295–304. https://doi.org/10.4236/ojf.2016.6402

Leff, E. (2003). La ecología política en América Latina: un campo en construcción. *Sociedade e Estado, 18*(1–2), 17–40. https://doi.org/10.1590/S0102-69922003000100003

Lindeman, R. (1942). The trophic-dynamic aspect of ecology. *Ecology, 23*, 399–418.

Lupasco, S. (1963). Las tres materias. Buenos Aires: Sudamericana.

Luthar, S., & Cushing, G. (1999). The construct of resilience: Implications for interventions and social policy. *Development and Psychopathology, 26*(2), 353–372.

Margalef, R. (1963). On certain unifying principles in ecology. *The American Naturalists, 25*(897), 357–374.

Martínez Miguélez, M. (2011). El paradigma sistémico, la complejidad y la transdisciplinariedad como bases epistémicas de la investigación cualitativa. *Revista REDHECS, 11*, 6–27.

May, R.M. (1992). How many species inhabit the Earth? *Scientific American,* 267(4), 42–48.

Morin, E. (1981). *La naturaleza de la naturaleza.* Ediciones Cátedra.

Morin, E. (1994). La noción de sujeto. En D. Schnitman F., Nuevos Paradigmas, cultura y subjetividad, pp. 67–85. Buenos Aires: Paidos

Morin, E. (2003). Educação e Complexidade: Os Sete Saberes e Outros Ensaios. São Paulo: Cortez.

Murphy, P. C. (2005). *What A Book Can Do: The Publication and Reception of Silent Spring*, University of Massachusetts.

National Aeronautics and Space Administration (NASA). (2017). Global Climate Change: Vital Signs of the Planet. Available at: https://climate.nasa.gov/ (Accessed 15 June 2026).

Odum, E. (1953). *Fundamentals of ecology.* W.B. Saunders Company.

Organización de la Naciones Unidas. (1987). Informe de la comisión mundial sobre el medio ambiente y el desarrollo. Tomado de: http://www.ecominga.uqam.ca/PDF/BIBLIOGRAPHIE/GUIDE_LECTURE_1/CMMAD-Informe-Comision-Brundtland-sobre-Medio-Ambiente-Desarrollo.pdf

Organización de la Naciones Unidas. (2005). Conferencia Mundial sobre la Reducción de los Desastres, Kobe: Hyogo, Japón, 18 a 22 de enero de 2005.

Organización de las Naciones Unidas. (1992). Conferencia de las Naciones Unidas sobre el medio ambiente y el desarrollo, Río de Janeiro.

Osorio García, S. (2011). El pensamiento complejo y la transdisciplinariedad: fenómenos emergentes de una nueva racionalidad. *Revista Facultad de Ciencias Económicas, 20*(1), 269–291. https://doi.org/10.18359/rfce.2196

Otero Carvajal, L. (1992). De Darwin a Einstein. En Bahamonde Magro, A. La época del imperialismo (pág. cap I). Barcelona: Planeta.

Pauli, W. (1925). Über den Zusammenhang des Abschlusses der Elektronengruppen im Atom mit der Komplexstruktur der Spektren. Zeitschrift für Physik, 31, 765–783.

Pombo, O. (2013). Epistemología de la interdisciplinariedad. La construcción de un nuevo modelo de comprensión. *Interdisciplinar I, 1*, 21–50.

Prigogine, I. (1996). El fin de las certidumbres. Barcelona-Santiago de Chile: Andrés Bello, Print.

Puig, G., & Rubio, J. (2013). *Manual de resiliencia aplicada.* Gedesia editorial.

Rebotier, J., López Peláez, J., & Pigeon, P. (2013). Las paradojas de la resiliencia: miradas cruzadas entre Colombia y Francia. *Territorios*, 127–145.

Rodríguez, G. A. V., Lucho-Constantino, C., Olivares, C. C., & Hernández, I. B. (2014). Esbozo histórico de las Ciencias Ambientales I. De Hipócrates a primavera silenciosa. *PäDi Boletín Científico De Ciencias Básicas e Ingenierías Del ICBI, 2*(3). https://doi.org/10.29057/icbi.v2i3.528

Rumer, B., & Ryvkin, M. (1980). Thermodynamics, statistical physics, and kinetics. Moscú: MIR Publishers.
Rutter, M. (1991). Resilience: Some conceptual considerations. *Journal of Adolescent Health, 14*(8), 626–631.
Stocker, T. (2014). *Climate change 2013: The physical science basis: Working Group I contribution to the Fifth assessment report of the Intergovernmental Panel on Climate Change*. Cambridge University Press.
Torres, C. (2016). Reflexiones alrededor de la epistemología ambiental. *Revista de estudios sociales, 58*, 39–51. Tomado de https://dialnet.unirioja.es/servlet/articulo?codigo=5908040
Trujillo García, S. (2011). Resiliencia: ¿Proceso o capacidad? Una lectura crítica del concepto de resiliencia.
UNESCO. La Organización de las Naciones Unidas para la Educación, la Ciencia y la Cultura 1969. Actas de la conferencia general de Paris de 1968, 15.a reunión. Place de Fontenay, Paris 7e. 1969 CFS.69/VI.lS/S.
UNESCO. (2002). UNESCO Universal Declaration on Cultural Diversity. Paris: United Nations Educational, Scientific and Cultural Organization.
UNISDR. (2015). Marco de Sendai para la Reducción del Riesgo de Desastres 2015–2030. Tomado de: https://www.unisdr.org/files/43291_spanishsendaiframeworkfordisasterri.pdf
United Nations (UN). (1945). Charter of the United Nations. San Francisco: United Nations.
Weil, S. (2000). The Need for Roots: Prelude to a Declaration of Duties towards Mankind. London: Routledge.
World Health Organization (WHO). (2016). Preventing Disease Through Healthy Environments: A Global Assessment of the Burden of Disease from Environmental Risks. Geneva: World Health Organization.

Chapter 3
A Conceptual Proposal of Socio-Ecological Resilience Indicators for Rural Populations

María Alejandra Taborda Caro, Rubén Sepúlveda, Carmen Ortega Otero Socióloga, Ángela María Moreno Barros, Rafael Prado, Jaime Muñoz Berrio, and Yamith Cuello

Abstract This work carries out a theoretical review, and conceptually proposes a battery of indicators, for the construction of a model of management of sustainability and participatory work with rural populations and indigenous peoples. It is also a pedagogical instrument, which can be used as input for the development of socio-environmental investment projects, to counteract the climate emergency or give way to community initiatives, which improve the scenarios for decision-making, on the action's conservation of the environment, from the perspective of the governance of the territory and the safeguarding of ancestral knowledge linked to ecosystems. This work brings three models of interpretation from the Social Sciences considering the society and the ecological environment: (i) focus on the dissolution of the community by the individual through Historical Interpellation; (ii) proposes an analysis from a perspective of cultural Anthropology, focused on Resilience, understanding it as "communality" or local autonomy and cultural knowledge; and finally, (iii) a proposal for indicators of socio-ecological resilience.

Keywords Socio-ecological resilience · Communality · Local autonomy · Sustainability · Traditional knowledge

M. A. T. Caro (✉) · C. O. O. Socióloga · Á. M. M. Barros
Unicórdoba, Córdoba, Colombia
e-mail: mataborda@correo.unicordoba.edu.co

R. Sepúlveda
Universidad Pontificia Bolivariana, Montería, Colombia

R. Prado
UNESCO Chair on Sustainability—Polytechnic University of Catalonia, Barcelona, Spain

J. M. Berrio · Y. Cuello
SIC Unicórdoba, Córdoba, Colombia

J. Morato et al. (eds.), *Intangible Cultural Heritage and Traditional Knowledge Related to Urban Resilience*, Resilient Cities, https://doi.org/10.1007/978-3-032-12713-6_3

3.1 Local-Global Context

The emerging discourses of capitalist modernity have modified the larger society and rural communities, as they propose a model that does not allow any space for the exchange of ancestral and traditional knowledge and practices or empirical knowledge. It is therefore necessary to enact spaces for dialogue and achieve, with *in situ* interventions such as actions to safeguard the identity connected to traditional knowledge of indigenous and traditional peoples, in the face of the overwhelming force of economic development. Thus, in other words, the community knowledge, related to biocultural heritage *e.g.*, must be articulated with the system based on resilience as a system of practices of resistance and self-organization, which mitigate the effects of climate change and loss of cultural heritage.

In this logic, the concept of "communality" is configured as an axis of alternative strategies against the mechanical hegemonic model, emanating from capitalist economic rationality. The concept of communality focuses on revitalizing discourses to transform the ecological and social fields, with propositional action for the dissemination of sustainable practices, which allows resilient responses for men and women of rurality affected by climate change around the world, to be activated through their biocultural heritage.

In this perspective, bio-cultural heritage will then be the mobilizing engine of a resilient system; due to its historical-ecological awareness and because it comes from a community *ethos*—owing to the culture and tradition of the peoples—made up of support networks, local productive forces, solidarity economies and traditional knowledge; all intrinsically sustainable. Nowadays, the need for heritage knowledge and socio-ecological practices arises as community strategies to counteract the imbalance at different scales from local to global.

3.2 The Historical Interpellation

The conclusion of the twentieth century occurred amidst continual political, social, and cultural transformations. As the product of two world wars, advances in science, technology and communications emerged, such as the invention of antibiotics and biotechnology, which dominated the landscape of modern science, with the purpose of improving the living conditions of the population. In contrast to the above, the most egregious extermination practices of recent centuries also occurred (Hobsbawm, 2007, 2013).

From the environmental perspective, there was the proliferation of the nuclear industry, the commercialization of new chemical products and the extractive industry that represents a quarter of the world's gross domestic product (GDP) of 81 countries, according to the World Bank (Banco Mundial, 2018). A practice that represents around 30% of its resources in Africa. The above data, by itself, does not account for the close link between the extractive industry and the polluting events with the greatest presence on the planet.

This has led to excessive global warming, the best-known effect of which is climate change or climate crisis. The uncertainty, the ambiguous and incomprehensible, became possibilities for scientific and social interpretation, previously distorted by international politics. The observations of the contemporaneity show us that the changes materialize with the modification of the temperatures, which in a systemic way will alter the rainy and windy seasons. Additionally, the production of the crops was modified and the life rhythms of various species, among others migratory birds, began to collapse on their journeys (Kearns, 2010; Judt, 2011).

In the cultural field, a group of practices emerged, which have become adaptive responses of a society, and which express a univocal relationship with the logics with which the environment is viewed: abiotic, biotic and anthropic, as a satisfaction to various human needs (Leff, 2000 and 2001). For others, culture is consolidated through the use of natural resources in peasant-indigenous productive systems, and determines their management and use, in addition to their forms of exploitation, in such a way that it is a social resource capable of being used. Destructively or rationally, of getting lost or developing, from other more subjective points of view.

Additionally, a cultural *ethos* constituted of rationalities and environmental practices is articulated in a duality, where the trends on consumption widen the gap, and the decrease of a multitude of plant and animal species, in line with the deterioration caused due to energy reserves, without reaching a global consensus on viable clean alternatives that mitigate the consequences of an extraction predator and unsustainable model.

Authors such as Elizabeth Kolbert in the book "The Sixth Extinction" affirm that the decrease in natural biodiversity is directly linked to the loss of cultural diversity (Kolbert 2011). The impact caused in the last century has generated a great mass extinction of animal and plant life, modifying man's behaviour with the environment. In this sense, we can affirm that biodiversity is in grave danger and unprecedented species extinction rates accelerating in dangerous decline.[1]

Following the United Nations Environment Programme (UNEP) reports the "biodiversity loss is caused by a range of different pressures, from habitat loss due to agricultural expansion, pollution or desertification to invasive alien species and climate change. Impacts of biodiversity loss are being felt across all Earth's major biomes and the cost of inaction is escalating, demanding increased investment in conservation and sustainable use on a global scale."[2]

---

[1] UNITED NATIONS. UN Report: Nature's Dangerous Decline 'Unprecedented'; Species Extinction Rates 'Accelerating': "By the Numbers—Key Statistics and Facts from the Report". <https://www.un.org/sustainabledevelopment/blog/2019/05/nature-decline-unprecedented-report/>

[2] UNEP GLOBAL BIODIVERSITY LOSS: SAFEGUARDING BIODIVERSITY FROM AN UNPRECEDENTED RANGE OF THREATS—SDG POLICY BRIEF NO. 003

United Nations Environment Programme (2018–05) <https://wedocs.unep.org/bitstream/handle/20.500.11822/25439/SDG_Brief_003_Biodiversity_201805.pdf?sequence=1&isAllowed=y>

The unlimited consumption of resources generates two impacts, both linked in a single dynamic: habits and customs. For example, the peasants of organic agriculture, which still exist in tropical climates, have customs for pruning, planting and harvesting linked to the lunar calendar, in addition to the chirping of birds and the smell of the wind, this type of knowledge disappears quickly, as the environment that allows it changes, without, very often, people being aware of the consequence of this loss.

With the structural changes of a planetary nature, we have even come to think that we are already located in a new era that the biologist Eugene Stoermer and the Dutch meteorologist and Nobel laureate Paul Crutzen called the Anthropocene. Since then, communities of biologists, geographers and geologists have been in a rush because their periodization forces an arbitrary rupture in what is a lasting process of human alterations in ecosystems.

Studies on socio-environmental processes insist on the constant changing of global humanity (Finney & Edwards, 2015). The authors argue that geological time based on a "change of step" is the result of heterogeneous activities rooted in socio-political contexts, even more complex networks, with environmental transformations at multiple scales. Thus, the periodization of the Anthropocene, what can be called the "Anthropocene split", obscures rather than clarifies the understanding of human-environmental relationships (Bauer & Ellis, 2018).

We certainly live on a limited planet, which requires the wise use of natural resources. Urgent cultural and behaviour changes are necessary calling to involve states of consciousness and ethical views, such as the recovery of the values of austerity and the appreciation for durability and the greatest possible use of the objects that accompany us daily (León, 2008). A look back at the trades, at the simple ways of living, at the need to understand ourselves as part of a great living universe, where we have the responsibility to validate our acts as thinking beings, with the capacity for abstract thought and with the possibility to reflect on the daily life between us and the ecosystems.

In 1972, with the publication of the Meadows report "The limits to growth" were addressed the insistence on the evident infeasibility of the permanent growth of the population and its trends and the need to find in recycling, the return to the traditional cycles of the materials, and their stocks to allow the functioning of natural processes (Maletta, 2011). Today they are looking for the management of the energy crisis to force transport to be minimal and the products and resources used to be close. An economy settled in the near future, allows the environment to combine with the less vulnerable communities since it allows them greater control and independence of the decisions that are made in the global power centres.

Future economic sustainability in southern countries requires, beyond waste management, managing democratic governance of a regional nature, making use of local labour force and materials, within the framework of stable ecological models. Additionally, other issues such as the participation of women and everything related to gender, become determinants of public policies and dialogues of knowledge. This occurs, according to Mojica (2013, p. 78), "because women have been very close to the material conditions of subsistence, they have developed jobs and skills that make them more adapted to walk towards sustainability".[3]

In relation to the positions on the impact of the environmental crisis, Mercado and Ruiz (2006) show that there are two approaches at a global level that have been appearing since the 90's: the first, within emerging or southern countries, which are trying to explain how the conditions of extreme poverty and survival lead to an over-exploitation of resources and the other, affirms that the exploitation of natural resources is an irreconcilable product of the capitalist system. These confrontations, the author explains, have to do with the North-South vision of development. We would thus be faced with two paradigms, one of a rationalist line derived from the Enlightenment: technology-modernity-civilization and the role of new non-polluting technologies and, the other, from an anthropological perspective of the Latin American paradigm, which vindicates the tradition of indigenous peoples, which defends the dialogue of knowledge.

In addition to the paradigms set out above, the environmental crisis was configured with the risk theory shown by sociologists Beck and Giddens, which allowed us to understand how the sophistication of technology, and the use of energy sources, made natural risk become re-assessable as a culture of prevention and environmental intervention, which could be carried out at different scales, capable of recognizing in society, the impact and effects caused by human action (Maletta, 2011).

The "social" as category has been shaping itself, in the light of the new explanatory and problematic paradigms that have emerged in a dizzying way in our environment: in architecture, furniture, food, aesthetics, and religion. One of the problems in facing the contemporary global social crisis, Fernández (2014) affirms, is our difficulty in understanding it in its complexity and roots. In view of this, we have the need to rethink traditional academic disciplines, since much of their complex thinking, methods and actions could not be framed in the narrow fields of specific or monodisciplinary academic knowledge (López and Ospina, 2008).

In the difficult task of understanding society, theoretical paradigms were created, which we can synthesize as follows: Today, society is not defined as a national space or as the aggregation of human individuals (Rigby, 2011); We can understand social reality as the assembly that decentres the human being and that is called posthumanism or post-social (García, 2010).

The study of human beings is currently understood as an individual-public with an overexposure of their privacy (Becerra, 2015), as a consumer society, (Baudrillard, 2009; Virilio, 2012), as a remote-controlled society with a preponderance of what is visible over the intelligible (Sartori, 2009).

[3] As the original: "las mujeres por haber estado muy cercanas a las condiciones materiales de subsistencia han desarrollado trabajos y habilidades que les hacen estar más adaptadas para caminar hacia la sostenibilidad", the translation is free made by the authors of this work.

3.3 Resilience: Human—Environ Mental Relations System

The concept of resilience emerged within the social sciences from the 80s, especially incorporated by psychology to show people capable of staying psychologically healthy despite living in high-risk contexts, such as environments of poverty and multi-problem families, situations of prolonged stress, in detention centres, among other factors. The concept of community resilience had its emergence later on, applied to different social situations (earthquakes, floods, cyclones, famines, civil wars, guerrillas, repression, etc.), events that appropriate survival strategies, organization schemes, commitments, bonding, and other experiences to establish new social and labour ties (Arciniega, 2013; Brangwyn & Hopkins, 2010; Marzana et al., 2013).

Individual transformations affect collective impacts, these aim to empower individuals and grassroots social organizations to change adverse realities with multicausal views of the problems. Some theories have already begun to visualize this impact as the Collective Efficacy developed by Sampson, which highlights the importance of an organized community as a potential to control the dynamics of security and coexistence of the environments (Sampson & Laub, 1996) *apud* (Valenzuela, 2012).

Community resilience "is the collective condition to overcome disasters and massive situations of adversity and build on them." This perspective emphasizes people's capacities to adapt to change, the ability to learn, creativity, orientation towards the future, strengths and opportunities. One of the most important actions in this dimension focuses on the possibility of informing groups about risks and emergency situations with the aim of preparing with preventive and remedial actions (Acinas, 2007). The collective is presented as the belief of being able to act together and achieve the expected effect.

Due to the intrinsic relationship between the human being with nature and the *ethos* that communities develop, socio-ecological resilience will be included in community identity and organization, where the components of communality, local autonomy, local knowledge and capacities are highlighted. Key elements in the conservation of communal and environmental values, in addition to the maintenance and reaffirmation of unity based on the collective memory of the peoples compose this concept.

The first element in the process of social-ecological resilience which we can define as communality is the movement of the original peoples of Oaxaca-Mexico, which arises with the purpose of valuing and protecting community knowledge and practices against government repression from the 80's. It was conceived as a way of life and communal organization differs from that managed in the West.

Rendón (2004) argues that communality is a "way of naming and understanding Indian collectivism [...]" (p. 367), and although it is a word that originates from indigenous peoples, it also has a place in non-aboriginal communities but that have a participation and reciprocity around a territory (Rendón, 2004). The term is adapted to community contexts, where social, productive, organizational relationships have the communal as essential (Diaz, 2004; Martinez, 2015). For his part, Toledo (2015) adds that the communal values of cooperation, solidarity, altruism and reciprocity, and their relationship with the environment, is a determining factor of communality, being the necessary elements to face the destructive forces of neoliberalism.[4]

"(…) We are communality, the opposite of individuality, we are communal territory, not private property; we are compartment, not competition; we are polytheism, not monotheism. We are exchange, not business; diversity, not equality, although in the name of equality we are also oppressed. We are interdependent, not free. We have authorities, not monarchs. Just as the imperial forces have relied on law and violence to subdue us, we rely on law and harmony to reply, to announce what we want and wish to be (…)".

Thus, communality is linked to the resistance of peoples to social and environmental threats. For this, the development of lifestyles and a collective mentality contributes to the strengthening of the organization and a communal structure (Maldonado, 2015) with the ability to face the challenges of today's world, being a social tool in the resilience of the communities and their environment.

The indigenous Mixe Floriberto Díaz, highlights some elements in the communal organization: "Earth, as Mother and Territory; consensus in assembly, for decision making; Free Service, as an exercise of authority; Collective work, as an act of recreation; Rites and Ceremonies, as an expression of the communal gift" (p. 40). Based on these elements, Fuente (2012) points out some general attributes in the institutional organization of native peoples and local communities, which strengthen interpersonal ties in the territory, these are: Organization of community work, community territorial possession, construction of identity cultural and worldview.

"The organization of community work […] linked to assessments such as the development of local prestige or commitments imposed from the community" to continue belonging to it "[…]; Community territorial possession […] basis for transformation, expression of specific knowledge on the use of natural resources […] Construction of cultural identity […] complex notion that feeds on various aspects, including territorial support and the spiritual religious representations of the past in the present and the attachment to the earth […] The worldview is the manifestations about the cultural perception of nature […]".[5]

The attributes described contribute to having an appropriation of the natural environment in accordance with communal life and detached from the hegemonic positions of Western society, thereby contributing to the approach in the relationship between man, society and ecosystem.

The second element of socio-ecological resilience, local autonomy, is oriented to control and management in the territories without external interference, determining for the consolidation of governance and the development of projects with a local focus.

For Hernández and Santiago (2014), this autonomy is represented in "the freedom of the communities to own, control and manage their territories, to regulate their

[4] Free translation from the original

[5] Free translation from the original

political, economic, social and cultural life [...]" (p. 4),[6] making use of local customs and knowledge in the consolidation of organizational parameters, this without not opposing the state institutionally.

According to Escalera and Ruiz (2011), the self-determination of communities is necessary in the sustainability of an environment, since through relations of power and autonomy from local settings, citizen processes are established for the defence of the territory, supported by the protection of their home, their food and economic sustenance, their community and cultural scene, their heritage. In this regard, Chapín apud Boege (2017) highlights the importance of autonomy and local identity for the development of resilient processes.

"[…] Adaptation, resistance and renewal in the face of these adverse situations on the part of local cultures and indigenous peoples occurs when they manage to maintain their autonomy in decision-making, and—we can add—retain and develop control culture" (p. 56).[7]

For Boege (2017), autonomic processes are duly related to biocultural heritage, insofar as individual and/or group aspects such as communal identity and socio-ecological imaginary lead to collective empowerment and the redefinition of the local organizational structure, achieving face state impositions and globalization. The same author defends the posture that the society-nature relationship should be assumed in cultural rights, guarantees autonomy, food sovereignty, communal advancement and the enjoyment of biocultural heritage (Boege, 2017).

Likewise, Hernández and Santiago (2014) highlight the value of autonomy in community organization, featuring collective values of indigenous peoples in Mexico, which constitute systems of resilience and community development in contexts where globalization threatens self-determination. And for this, the rescue of identity from daily activities of social and productive order has been important.

The authors identify communal aspects, found in Mexico, that are necessary in autonomous construction, such as: "Recovery of the environment, Communal property, Permanence of the rural population, Food self-sufficiency, Better welfare conditions, Recovery of values: equality, justice, democracy, and solidarity" (p. 3).[8]

The construction of autonomy leads to the production of the community *ethos* (Fuente, 2012), necessary for the resistance and advancement of peoples through local control of socio-productive actions, based on the diversity of institutions and the natural environment. Fuente rescues local autonomy processes that allow the development of alliances in communities, where the citizen's decision will be focused on sustainability and good living in the territory, based on community productive forces, market diversification, support networks and sovereignty food. Noting in this regard:

"The political-culture formation expressed in relations of subordination, resistance or greater political autonomy from the government […]; The development of the community productive forces expresses the development and technological

[6] Free translation from the original

[7] Free translation from the original

[8] Free translation from the original

appropriation, productive diversification (ecotourism companies, water purification plants, community stores, etc.) […]; Market diversification […] Process based on the premise of a necessary search for alternatives in the spheres of exchange and circulation […]; The formation of support networks […] deployment of activities with local social and productive organizations […]; Food sovereignty in which the strategies of production, supply, intensity of use are defined "(para. 40).[9]

Communality and local autonomy, being supported by their own ideas, judgments and reasoning, take for granted an element of resilience focused on local knowledge and capacities, where cognitive aspects and cultural heritage contribute to assuming the adversities present in a community from the development of adaptive skills in the face of change. Escalera and Ruiz (2011) argue that local knowledge about culture and the environment […] guarantees the feeding of the socio-ecological memory as a source of innovation and novelty […]" (p. 121) in situations of crisis and disturbance.

Nevertheless, López and Limón (2017) argue that cultural knowledge is the result of the social fabric built with the history of a given territory. This knowledge is based on rootedness and permanence, they are a source of resilience to the extent that they occur over the years and define the behaviours of a community in its relationship with the environment, which Revilla, Martín and De Castro (2016) called *"baggage of the local population"* in free translation from Spanish, meaning the communal collective knowhow and knowledge, evidenced in the use of seeds, construction of houses, distribution of land and community organization.

Although transformations in the environment have repercussions on the development of local knowledge and on the identity of the peoples, communal life allows the conservation of a collective memory, fundamental in the feedback of knowledge linked to the social capacities historically developed in the community (Melillo et al., 2004). According to López and Limón (2017), local knowledge and capacities generate information with which the community resists disturbances and advances according to the collective desire, the authors highlight the resilient principles of: collective knowledge, collective cohesion and critical thinking collective, describing the following:

"Collective knowledge […] allows for the accumulation of meanings and social interpretations of a certain culture […]; Collective cohesion […] contributes to maintaining a shared future projection that fosters hope […] actions organized from and for common interests […]; Collective critical thinking […] allows us to redefine the violent events suffered, by analysing the contexts in which they occurred, re-dimensioning the causes and consequences, as well as the responsibilities of the participants "(pp. 5–7).[10]

---

[9] Free translation from the original

[10] Free translation from the original

3.4 An Approach to the Socioecological Resilience System Indicators

The resilience system is associated with sociocultural and organizational processes carried out within the community (Fuente, 2012). The development of these processes is subject to evaluation and monitoring in order to establish fundamental categories of socioecological systems, in the face of situations of change in the environment. For this reason, it is necessary to have a series of indicators built for the study of the sustainability of the territories, based on social values, self-management, adaptability to changes in the territories, the struggle for cultural autonomy, and elements of the biocultural heritage of the peoples.

Indicators as a methodological tool and result of survey methods for monitoring socioecological processes and to provide information for progress in decision-making, allow for the evaluation of the actions developed, contribute to improving management, generate commitments in the actors involved and contribute to accountability (CEPAL, 2009). In socioecological contexts, the indicators provide first-hand information and explain the transformations presented in a given space and time through the analysis and relationship of social, cultural and ecological variables.

Environmental conservation actions have required the role of indicators to measure the anthropic impact on the environment. Initially, only indicators of a biological nature were developed and framed in the physical restoration of the territory, then with theories of sustainable development since the Brundtland Report focused on reducing poverty, social indicators took their place in the scientific field, and later in view of the need to understand the complex relationship "Intervention-natural environment", the socio-environmental indicators were produced that contribute to the measurement of access and control of ecosystem services and their link with economic and socio-cultural aspects (CEPAL and PNUD, 2004).

3.5 Conceptual Proposal of Socioecological Indicators Systems for the Sustainability of Rural Populations

The identified categories of communality, local autonomy, local knowledge and capacities, constitute a route for the construction of socioecological indicators of a territory. With this tool, it is possible to evaluate the resilient capacities of adaptation, mitigation and transformation, all based on the sociocultural and environmental diversity of the territories. The conceptual proposal of indicators in this work is also based on the approach to rural populations, assuming socioecological contexts and realities.

Thus, the task of monitoring resilience in rural populations includes four groups of indicators, initially those referring to communality, their objective is to evaluate the collectivism of communities based on the conservation of cultural and ecological diversity. For the construction of this group of indicators, aspects of community coexistence are taken into account within the relationships of enjoyment and use of ecosystem services. Including in communality concept some collective, organizational elements such as: (i) Organization of community work, (ii) Community territorial possession, (iii) Construction of cultural identity and (iv) Cosmovision (Table 3.1).

A second round of socioecological indicators are placed in the category of local autonomy, these are intended to assess the self-determination of traditional, indigenous and rural peoples, freedom in making decisions of a social, productive and ecological nature. Among the main obstacles to local autonomy, is external intervention in political affairs and dependence on other communities in production and food issues. For the elaboration of these indicators, elements of local governance and control are taken into account in productive matters, trade and satisfaction of basic needs, these contain aspects related to: Organizational capacities, strengthening of local processes, political-cultural training, community productive forces, market diversification, support networks, and food and alimentary sovereignty/autonomy (Table 3.2).

A third set of indicators are framed in local knowledge and capacities, where the evaluation of their own knowledge and social skills in the face of change are the central object of these indicators. Therefore, it is especially pertinent the observation focused on the loss of traditional knowledge related to productive and cultural activities, the collective interest in situations of change in the territory. In reference to local knowledge and capacities, López and Limón (2017), point out characteristic elements such as: collective knowledge, collective cohesion and collective critical thinking (Table 3.3).

The registered indicators are an initiative for the evaluation of the cultural and ecological diversity of the traditional populations from "La Mojana" area, Ancestral territory of the Zenú ethnic group in the Colombian Caribbean region. Starting from a participatory analysis, where the collective organization, characteristics of identity,

Table 3.1 Communality indicators

Category	Indicators	Descriptions
Communality	Participation in community activities	Development of communal works of collective benefit
	Common land tenure	Collective spaces for socio-productive projects
	Traditional and customary community domestic law-rules	Recognition of political-cultural identity in communal management
	Festivities and ceremonies related to cultural and environmental identity	Development of cultural activities that strengthen the identity of the population connected to the territory
	Conservation of sacred sites	Landmarks and key territories for the preservation of the worldview and communal rites

Table 3.2 Local autonomy indicators

Category	Indicators	Descriptions
Local and regional autonomy	Local management in decision making	Presence of the community in decision-making related to the dependent ecosystem
	Community coordination and cooperation mechanisms	Actions and collective disposition of support, stand out from adversity collectively
	Presence of productive associations in the community	The operation of farmers and/or fishing associations that strengthen local productivity
	Traditional forms of government	Management and conservation of a local government structure
	Food and alimentary sovereignty/autonomy.	Satisfaction of food needs from local production processes (orchards, ponds)
	Variety of agricultural production	The community obtains a diversity of agricultural products and ecosystem services

Table 3.3 Local knowledge and capacities indicators

Category	Indicators	Descriptions
Local knowledge and capacities	Seeds bank	Local knowledge regarding the conservation of creole (autochthon) seeds
	Storage of supplies and food	Store food to solve periods of low production
	Plague and illness management	Solution of pests and diseases in people and crops based on local knowledge
	Expectation of progress and collective change	Collective hope life derived from actions and projects of common interest
	Responsiveness to risk situations	Community critical knowledge derived from crisis situations

autonomy in the territory and local knowledge produced in the daily relationship with the environment were taken into account, all of the above being determining elements in the development of the resilience processes.

Currently in La Mojana, globalization, massive tourism, overexploitation of services and other factors have been permeating the identity of the peoples. This has motivated us to start evaluating the resilience of this territory from the biocultural heritage, its ability to cope with climate change and external threats from the

Table 3.4 Socio-spatial relationships indicators

Category	Indicators	Descriptions
Socio-spatial relationships	Land use	Knowledge based on Forest cover and recognition of biodiversity
	Knowhow of previous conservation projects	Programs (at the time of this study) with the aim of safeguarding ancestral knowledge associated with ecosystems
	Community and social start-up projects.	Identification of community social processes, life history methodologies and mapping of actors
	Description of government institutions	Mapping of actions and projects of common interest, proposed by local-national governments
	Decision making at the community level	Capacity for cohesion and execution of proposals.
	Monitoring activities in the territory with local stakeholders and actors	To know the experience of the community with the products obtained

conservation of ways of life typical of the Zenú people amphibious territory. The purpose is that the indicators contribute to the socio-ecological analysis in similar contexts (Table 3.4).

3.6 Conclusions

The proposal to build indicator instruments for the evaluation and monitoring of socioecological resilience through the Human Sciences, allows for spaces for generation for dialogue of knowledge and joint thinking about mitigation actions for the effects of climate change at the local to global scale.

Rural and/or indigenous peoples and traditional way of life populations worldwide, developed adaptive strategies and ways of relating through their culture with the environment, which allowed for infinite creativity in the use and exploitation of local resources. Access to resources is proportional to the supply of ecosystem services, as opposed to behaviours of overexploitation and alteration of the environment. For this purpose, monitoring human relationships, traditions, the worldview of peoples, allows for environmental references of the territory based on communal uses and practices.

The society-ecosystem relationship within the new perspective of the study of sustainability based on the safeguarding of Biocultural Heritage, represents the most important challenge for future research from trans-disciplinary nature, with a service approach to these traditional communities taking into account the fast deterioration of the worldwide biodiversity.

This conceptual proposal of indicators for the monitoring and evaluation of the socio-ecological system is adaptable to any context where the conservation of cultural and environmental values is promoted as a local resistance strategy against extractive practices and/or over-exploitation, forces that cause ecocides, ethnocides as a result of environmental impact and imbalance.

Similarly, the indicators brought in this work were developed within assimilable objectives, which contribute to social progress, together with ecological stability and the development of adaptable and innovative participatory coexistence models, to assume the social and ecological adversities of the new scenario. Climate emergency.

References

Acinas, P. (2007). Información a la población en situaciones de emergencia y riesgo colectivo. *Intervención psychosocial: Revista sobre igualdad y calidad de vida. 16* (3), pp. 303–322.

Arciniega, U. (2013). La perspectiva comunitaria de la resiliencia. *Psicología Política, 2013*(47), 7–18.

Mundial, B. (3 de Octubre de 2018). Industrias extractivas: Contexto. Disponible en: https://www.bancomundial.org/es/topic/extractiveindustries/overview#1.

Baudrillard, J. (2009). La sociedad de consumo sus mitos, sus estructuras. Madrid: Siglo XXI.

Bauer, A., & Ellis, C. (2018). The Anthropocene divide obscuring understanding of social-environmental change. *Current Anthropology 59*, (2). Disponible en: https://www.journals.uchicago.edu/doi/pdfplus/10.1086/697198.

Becerra. (2015). Intimidad y Privacidad en Entornos Digitales luego de la Reforma del Código Civil. Simposio Argentino de Informática y Derecho.

Boege, E. (2017). Biocultural heritage and the cultural rights of indigenous people, local communities and their equivalents.

Brangwyn, B. & Hopkins, R. (2010). Compendio de iniciativas de transición, Ediciones EcoHabitar, Teruel.

CEPAL. (2009). Curso-Taller para la construcción de indicadores de desempeño. Construcción de Indicadores de desempeño con base en la metodología de marco lógico.

CEPAL and PNUD. (2004). Diseño de un sistema de indicadores socio ambientales para el Distrito Capital de Bogotá.

Díaz, F. (2004). Comunidad y comunalidad. En Culturas populares e indígenas, pp. 365–373.

Escalera, J., & Ruiz, E. (2011). Resiliencia Socioecológica: aportaciones y retos desde la Antropología. Revista de Antropología Social. (En línea) 2011, 20, pp. 109–135. Disponible En: http://revistas.ucm.es/index.php/RASO/article/view/36264.

Fernández, T. (2014). Crisis y Estado de bienestar: las políticas sociales en la encrucijada.

Finney, S. & Edwards, L. (2015). The "Anthropocene" epoch: ¿Scientific decision or political statement? GSA Today, 26(3–4): 4–1.

Fuente, M. (2012). La comunalidad como base para la construcción de resiliencia social ante la crisis civilizatoria, Polis [En línea] 2012, 33. Publicado el 23 marzo 2013. URL: http://polis.revues.org/8495.

García, F. (2010). Argumentos para una Sociología posthumanista y postsocial. Athenea Digital. Revista de pensamiento e investigación social, (en línea) 2010, 19. Pp. 7–27. Disponible en: https://dialnet.unirioja.es/servlet/articulo?codigo=3687876.
Hernández, L. & Santiago, M. (2014). Los procesos de resiliencia en las comunidades campesinas de México. Conference: Séptimo Coloquio Interdisciplinario de Doctorado. Disponible:https:// www.researchgate.net/publication/268983074_Los_procesos_de_resiliencia_en_las_comuni dades_campesinas_de_Mexico
Hobsbawm, E. (2007). Guerra y paz en el siglo XXI Crítica,
Hobsbawm, E. (2013). Un tiempo de rupturas. Sociedad y cultura en el siglo XX, México, Crítica, 2013, p. 312.
Judt, T. (2011). Algo va mal. Bogotá: Distribuidora y Editora Aguilar, Altea, Taurus, Alfaguara S. A.
Kearns, C. (2010). Conservation of biodiversity. *Nature Education Knowledge, 1*(9), 7.
Kolbert, E. (2011). La era del hombre. Fecha de (Consulta 3 de 2020). Disponible en: http://www. nationalgeographic.com. es/2011/02/28/era_del_hombre.htm.
Leff, E. (2000). Pensar la Complejidad Ambiental. En: La Complejidad Ambiental. México: Siglo XXI Editores.
Leff, E. (2001). Transgénesis. En: Formación ambiental. 13, (28).
León, T. (2008). Tierra, agricultura y ambiente: ¿es el desarrollo una categoría de la dimensión ambiental o viceversa? Innovación y Ciencia, XV (3), 60–69.
Castaño, C.E.L, Ospina, G.A. (2008). Ecología histórica. Interacciones Sociedad–Ambiente a distintas escalas socio-temporales, pp. 267–272. Pereira, Colombia: Universidad Tecnológica de Pereira, Universidad del Cauca, sociedad de Arqueología de Pereira.
López, C. & Cano, M. (2008). Aportes de la Ecología Histórica a la Cultura Ambiental. En C. E.
López, F. & Limón, F. (2017). Componentes del proceso de resiliencia comunitaria: conocimientos culturales, capacidades sociales y estrategias organizativas. El proceso de resiliencia comunitaria. Psiencia.
Maldonado, B. (2015). Perspectivas de la comunalidad en los pueblos indígenas de Oaxaca. Bajo el Volcán [en línea] 15, (23), pp. 151–169. [Fecha de consulta: 27 de Junio de 2018] Disponible En: http://www.redalyc.org/articulo.oa?id=28643473009.
Maletta, H. (2011). Blog Bid: Hablemos de Cambio Climático. Disponible En: http://Blogs Iadb.Org/Cambioclimatico/2011/06/20/15oblación-Y-CambioClimatico-¿Estaban-Malthus-Y-El-Club-De-Roma-En-Lo-Cierto/
Martínez, J. (2015). Conocimiento y comunalidad. Bajo el Volcán [en línea], 15 (23): [Fecha de consulta: 2 de Febrero de 2020] Disponible en:<http://www.redalyc.org/articulo.oa?id=286434 73006> ISSN 8170–5642.
Marzana, D., Marta, E., & Mercuri, F. (2013). De la resiliencia individual a la resiliencia comunitaria. Revista Electrónica Interuniversitaria De Formación Del Profesorado, 16(3), pp. 11–32. doi:https://doi.org/10.6018/reifop.16.3.186961.
Masera, O., Astier, M., & López-Ridaura, S. (2000). Sustentabilidad y manejo de los recursos naturales. El marco de evaluación MESMIS. México.
Melillo, A., Suárez, E., & Rodríguez, D. (2004). Resiliencia y subjetividad. Los ciclos de la vida. Paidós. Bs. As.
Mercado, A & Ruiz, A. (2006). El concepto de las crisis ambientales en los teóricos de la sociedad del riesgo. Espacios Públicos [en línea] 2006, 9 [Fecha de consulta: 5 de Febrero de 2020] Disponible en:<http://www.redalyc.org/articulo.oa?id=67601813> ISSN 1665–8140.
Mojica, F. (2013). Los negocios del futuro en la era del conocimiento en el Gran Caribe. Memorias, 1er Congreso de Prospectiva y Estudios del futuro de la Cuenca del Caribe–Prospecta Caribe. Cartagena.
Rendón, J. (2004). La comunalidad o modo de vida comunal de los pueblos indios. México: Dirección General de Culturas Populares e Indígenas.
Revilla, J., Martín, P., & De Castro, C. (2016). Resiliencia y acción social: la construcción cultural y colectiva del afrontamiento de la crisis. XII Congreso Español de Sociología.

Rigby, S. (2011). Historia, discurso y el paradigma postsocial: ¿una revolución en historiografía? Anales De Historia Antigua, Medieval Y Moderna, 43. https://doi.org/10.34096/ahamm.v43.2598.
Sampson, R. J., & Laub, J. H. (1996). Socioeconomic achievement in the life course of disadvantaged men: Military service as a turning point, circa 1940–1965. *American Sociological Review*, 347–367.
Sartori, G. (2009). El Homo videns: La sociedad teledirigida. México: Taurus.
Toledo, V. (2015). Comunalidad: el poder subversivo de la cooperación. (En línea] Publicado el 27 de Octubre 2015, [consultado el 3 de Febrero 2020] URL: http://www.jornada.unam.mx/2015/10/27/opinion/016a1pol
Valenzuela, A. (2012). La eficacia colectiva como estrategia de control social del espacio barrial: evidencias desde Cuernavaca, México. Revista INVI versión On-line ISSN 0718–8358.
Virilio, P. (2012). La administración del miedo, pasos perdidos. Madrid: Barataria.

Chapter 4
Traditional Knowledge and Intangible Cultural Heritage for Adapting to Climate Change

Jordi Morato, Luis Diaz, Katarzyna Alicja Wieszczeczynska, Maria Pizzorni, Rafael Prado, María Alejandra Taborda Caro, Apolinar Figueroa, and Nicola Tollin

Abstract Climate change and environmental degradation are an existential threat to the world. To overcome these challenges, the world needs a new growth strategy that will transform the planet into a modern, resource-efficient and competitive integrated economy. The degree to which populations are vulnerable to hazards is not solely dependent upon proximity to the source of the threat or the physical nature of the hazard, because social factors also play a significant role in determining vulnerability. In every place, the hazard potential interacts with the underlying social fabric of the place to create the social vulnerability. Today, one of the main challenges for climate adaptation and mitigation is to integrate the culture of communities that live in a given habitat and specially the intangible cultural heritage (ICH). Therefore, new approaches and strategies to face climate change should encompass the review of past and existing cultural practices used by indigenous communities to face environmental problems. We can turn crisis into opportunities and making the transition just

J. Morato (✉)
Technical University of Catalunya, Terrassa, Spain
e-mail: jordi.morato@upc.edu

L. Diaz · R. Prado
Polytechnic University of Catalonia, Terrassa, Spain
e-mail: luis.david.diaz@upc.edu

K. A. Wieszczeczynska · M. Pizzorni
University of Southern Denmark, Odense, Denmark
e-mail: kawi@iti.sdu.dk

M. A. Taborda Caro
University of Córdoba, Montería, Colombia

A. Figueroa
University of Cauca, Popayán, Colombia
e-mail: apolinar@unicauca.edu.co

N. Tollin
SDU Civil and Architectural Engineering, University of Southern Denmark, Odense, Denmark
e-mail: nto@iti.sdu.dk

J. Morato et al. (eds.), *Intangible Cultural Heritage and Traditional Knowledge Related to Urban Resilience*, Resilient Cities, https://doi.org/10.1007/978-3-032-12713-6_4

and inclusive for all. However, our possibilities will be considerable improved if we recognize and understand all the different forms of cultural–traditional/vernacular/ancestral-practices and knowledge based in the evolution of societies living during centuries in symbiosis with nature. Many people have intelligently benefit from ecosystems and have managed to adapt to environmental disturbances for centuries with a symbiotic living using traditional ecological knowledge (TEK). With the aim of publicizing the best practices with the possibility of replicating in urban environments, a selection of 11 methods used by ancestral cultures for the management of their necessary resources for survival in a sustainable manner without generating negative impacts on the environment was carried out. The purpose of this selection is to recognize the importance of the practices used by ancient cultures in managing climate change, and to publicize their ability to replicate and adapt to current environments. This recognition is an initiative to demonstrate the technologies and practices that current governments can use to deal with the problems generated by the climate emergency (floods, droughts, among others).

Keywords Climate change adaptation · TEK · ICH · Social-fabric · Best practices

4.1 Introduction

During recent years, the level of uncertainty due to climate change has generated a notable increase in debates on climate change (Davidson et al., 2020) seeking strategies that allow increasing the resilience of societies. The increase in climate emergencies due to extreme climate events such as large-scale floods, record temperatures for long periods of time, among other problems has increased awareness globally, even reaching some governments, that have declared climate emergencies.

In order to counteract the extreme negative impact of human activities on earth, especially due to industrialization and urbanization processes, novel approaches are required that allow to face a rapid process of transformation at all scales and sectors. Action for transformation is the motto for many organizations that seek to face the transition towards a low carbon and resilient society.

Effective mitigation of losses from environmental hazards requires hazard identification with the assessment of all the hazards likely to affect a given place, and risk-reduction measures that are compatible across a multitude of hazards. However, the degree to which populations are vulnerable to hazards is not solely dependent upon proximity to the source of the threat or the physical nature of the hazard, because **social factors also play a significant role in determining vulnerability**.

In every place, the hazard potential interacts with the underlying **social fabric** of the place to create the social vulnerability (Cutter et al., 2000). The social fabric includes sociodemographic characteristics, perception of and experience with risks and hazards, and overall capacity to respond to hazards. Today, one of the **main challenges for climate adaptation and mitigation is to integrate the culture of**

communities that live in a given habitat (Hulme, 2009; Jackson et al., 2018), and specially the **intangible cultural heritage (ICH)**.

Therefore, **new approaches and strategies to face climate change should encompass the review of past and existing cultural practices used by indigenous communities** to face environmental problems (Bardsley, 2017).

It is no coincidence that the mass extinction of diversity on earth is closely following the displacement of indigenous peoples. Innovations from the indigenous world are being lost at the same pace of biodiversity, and their displacement remains in fact an unrecognized driver of species extinction.

This opening article will focus on the ICH linked to conservation, environmental governance and biodiversity issues. Every society with changing ecological systems, presents cultural adaptation process. Political actions on management of environmental resources, enrol as new trends that develop scenarios for human adaptation from the political economy, and this will require the use of all their cultural background.

A review of best practices and cases of traditional knowledge and ICH for adaptation to climate change is presented, with a detailed exposition of the 8 most relevant ones.

4.2 Traditional Knowledge and Adaptation Capacity for Climate Resilience

Climate change and environmental degradation are an existential threat to the world. To overcome these challenges, the world needs a new growth strategy that will transform the planet into a modern, resource-efficient and competitive integrated economy. For example, the **European Green Deal is defining a new strategy to make the EU's economy sustainable** with no net emissions of greenhouse gasses, an economic growth decoupled from resource use and with no person and no place left behind.

The European Green Deal and other transition plans are based on circular economy, nature-based-solutions, ecosystem services, water-energy-food NEXIS, sustainable development and other new concepts, but practiced during centuries for many societies.

We can turn crisis into opportunities and making the transition just and inclusive for all. However, **our possibilities will be considerable improved if we recognize and understand all the different forms of cultural–traditional/vernacular/ancestral-practices and knowledge based in the evolution of societies living during centuries in symbiosis with nature**.

For example, **water has to be seen in connection with "urban green" to lead to blue–green nature-based solutions (infrastructure)**. The built environment defined by the urban space use and green infrastructures should be planned in combination with these blue and green features, to allow the characteristics of the natural water balance, in terms of infiltration, retention, evapotranspiration and run-off.

The concept of nature-based solutions should therefore become a core principle in every urban planning process, spreading multi-purpose green infrastructure in our cities (Liquete et al., 2016) in order to make their benefits available everywhere.

Following with the example, today water is sourced, or '**withdrawn**', from streams, rivers, lakes, reservoirs, oceans, and groundwater aquifers as well as harvested directly as rainwater. Circular economy is the principle of decoupling economic growth and development from the consumption of finite resources; this requires a transition to renewable resources and less dependence on finite resources. Much like materials, **human-managed water is today looked at in a linear fashion, in a Take-Use-Discharge strategy. To manage water systems for long-term sustainability, addressing the projected global and local demands for water, a circular approach is relevant, timely, and achievable and offers opportunity for commercial advantage**.

The term **'cultural heritage'** has changed content considerably in recent decades, partially owing to the instruments developed by UNESCO. Cultural heritage does not end at monuments and collections of objects. It also includes traditions or living expressions inherited from our ancestors and passed on to our descendants, such as oral traditions, performing arts, social practices, rituals, festive events, knowledge and practices concerning nature and the universe or the knowledge and skills to produce traditional crafts.

The resolution of the General Assembly 6 of the Intergovernmental Committee of the Intangible Cultural Heritage Convention (ITH/16/6.GA/7), recognizes in the Chapter VI, **the role of intangible cultural heritage as a driver and guarantee of sustainable development. The importance of intangible cultural heritage is not the cultural manifestation itself but rather the wealth of knowledge and skills that is transmitted through it from one generation to the next**.

While modern societies were trying to conquer Nature in the name of progress, the **indigenous cultures** were working with it. In contrast to the homogeneity of the modern world, indigeneity is reframed as an evolutionary extension of life in symbiosis with nature (Watson, 2019). That cumulative body of multigenerational knowledge, practices and beliefs is usually known as Traditional Ecological Knowledge (TEK).

Many people have intelligently benefit from ecosystems and have managed to adapt to environmental disturbances for centuries with a symbiotic living using TEK. Developed in direct contact with nature, TEK is engineered to sustain rather than exploit resources, fostering symbiosis between species by making biodiversity the building block used to construct green technologies. Indigenous technologies advance Darwin's theory of "survival of the fittest" to "survival of the most symbiotic." (Darwin, 1859; Margulis, 1998).

Indigenous communities have created synergies with nature over time, learning to coexist in the different environments where they settled. Transmitted from generation to generation, knowledge about the management of natural resources have allowed them to overcome environmental problems and to face drought, flooding and other environmental events.

The concept of **radical indigenism** takes its name from the Latin derivation of the word radical, meaning root. A school of design by radical indigenism rebuilds an **understanding of indigenous culture to generate sustainable and climate resilient infrastructures**. Once hybridized and scaled, these indigenous technologies could offer a new path to exponentially shrink the ecological footprint of humankind (Watson, 2019).

For indigenous peoples, traditional ecological knowledge that is conveyed through mythology forms the foundation of a complex understanding of the natural world. TEK is explained through four interrelated levels of ecosystem management, defined by a model known as the **Knowledge-Practice-Belief Complex**. This model coined by ecologist Fikret Berkes, translates the complexity of Traditional Ecological Knowledge into a Western scientific framework (Berkes, 1999). The first and foundational level is the local knowledge of animals, plants, soils, and landscapes. The second level involves resource management, composed of local environmental knowledge, practices, tools and techniques; and involving an understanding of ecological processes and performances.

The third level involves community and social organization offering coordination, cooperation, and governance. The fourth and final level is that of worldview which involves religion, ethics, and general belief systems. Each aspect of this Complex is important in maintaining a symbiosis between humans and nature.

These **TEK systems are embedded in culture and climate and are inherently sustainable**, being both an everyday response for human survival and an extraordinary response to environmental disturbances, such as drought, flood and fires.

From TEK, a catalogue of sustainable, adaptable, and resilient technologies that are borne out of necessity can be recognized. These indigenous socio-ecological technical systems (Gallardo, 2016) are local, inexpensive, hand-made, and easily constructed.

An emergent term, combining 'lo-tech' and 'TEK,' Lo-TEK reconceives the idea that indigenous innovation is primitive and exists in isolation from technology (Watson, 2019). The primitive is now innovative, redefining a misconception of indigenous innovation as being original, authentic, complex and designed to work with complex ecosystems. Lo-TEK is a movement that prompts us to retrace our steps and reconsider the root of technological innovation (Watson, 2019).

As one of the most outstanding examples, **the Zenu society, between 800 BC and 1200 AD** (FAO, 2019)**, learned to live with flooding, turning the excess of water as a big opportunity**. The Zenu ancestral territory comprises the valleys of the Sinú and San Jorge rivers, within the departments of Cordoba and Sucre in Colombia. The inland delta formed by these rivers and the Cauca river, the Magdalena river and the Nechi river frequently flooded during the rainy season in the mountains from April to November, causing great inconvenience to the residents of the plains. Therefore, from 800 BC to 1200 AD onwards these people built a system of channels that enabled them to control the flooding and make large areas practical for habitation and agriculture. The system was expanded continually, covering 500,000 ha (FAO, 2019).

The Zenú dug channels named "quebrales" connected to the natural waterways or "caños" (Aguilera-Díaz, 2004) in an intricated network including natural depressions in the terrain where wetlands or "zapales" are formed. Perpendicular to these channels, smaller irrigation ditches were dug. The soil left by the excavations was used to build long artificial terraces, two to four meters high, used for raising crops and built their houses. During times of high water, the channels led the water to areas where crops were grown. When the water withdrew during drought season, the nutrient-rich sludge was used to enrich the land. This system of water management, a clear demonstration of the Water-Ecosystems-Energy-Food (WEEF) NEXUS, was used over a period of 1300 years and many channel patters are still detectable in the landscape (Fig. 4.1).

Another great example of adaptation to climate change through traditional practices is reflected in the indigenous communities of north-western Australia, the Miriwoong, that have been learning about the seasons from past generations (Leonard et al., 2013). Being based on information passed on from generation to generation, traditional knowledge about seasonal patterns can add valuable insights to modern studies. Miriwoong traditional owners are confident that the use of long-established knowledge combined with scientific efforts will have a positive impact on the understanding of climate and climate change. Therefore, the recovery and understanding of ancient ways of land management linked to living within the environment can be combined today with modern conservation techniques to improve local resilience.

Fig. 4.1 Aerial view of the Zenú hydraulic system today (Guartinaja, Ayapel, Colombia). *Source* Jordi Morató

Traditional knowledge is becoming more attractive to scientists, and the collection of data on the knowledge of indigenous peoples, such as those already mentioned, offers an insight into the dynamics of ecosystems at a local scale (Leonard et al., 2013). We must begin to look to the past to learn from the mistakes of our ancestors and from the ancestral practices that allowed indigenous peoples to adapt to climate variations.

But even more, we need to see ancestral knowledge as a process, rather than as content, learning to formulate dialogues between science and ancestral knowledge (Berkes, 2009). The combination of traditional ecological knowledge with modern technologies would allow the creation of practices that improve our adaptation to climate change and reduce the levels of uncertainty.

The Alinytjara Wilurara peoples, in Australia, have created strategies in the management of natural resources, using fire as a tool for the creation of areas conducive to agriculture, but specially to mitigate the risk of uncontrollable forest fires (Bardsley, 2017). Based on their observations, they create fires on a small scale and intensity, in the places with the highest biomass providing large amounts of fuel for the summer times, when the vegetation dries, in order to prevent large forest fires (Bardsley & Wiseman, 2016).

It is extremely indispensable to increase our understanding on traditional ecological knowledge or cultural heritage from different parts of the world. Large information databases, knowledge, experiences, cases and best practices on climate adaptation and natural resources management can be found and documented everywhere (Shaffril et al., 2020). A need of better and more environmentally sound resources management—water, soil, food, …—that can take advantage of the country's climatic conditions and offers a good investment solution, it is specially needed at urban scale in developing countries, because the capital for infrastructure in these countries is usually scarce (Arias & Brown, 2009).

Active participation with communities can result in the creation of strong ties with the peoples, allowing the dissemination and sharing of knowledge, practices and innovation in materials to create new systems that allow societies to grow.

While numbering only one-twentieth of the global population, indigenous lands cover one-fifth of the earth's surface and shelter a staggering four-fifths of its number of species (Raygorodetsky, 2018). Over one million indigenous peoples have been forcibly removed from their lands in the last hundred years by conservation legislation alone (Dowie, 2009). While indigenous people have survived climate challenges for thousands of years using the ancient mythologies of technology contained in TEK, they cannot survive this global displacement.

4.3 Legal Aspects for Traditional Knowledge and ICH Conservation and Recovery

In the VI.3.3 part of the Operational Directives for the Implementation of the Convention for the Safeguarding of Intangible Cultural Heritage (https://ich.unesco.org/en/directives), **on Community-based resilience to natural disasters and climate change**, the GA stated that all parties should endeavour to ensure recognition of, respect for and enhancement of knowledge and practices concerning geoscience, particularly the climate, and harness their potential to contribute to the reduction of risk, recovery from natural disasters, particularly through the strengthening of social cohesion and mitigation of climate change impacts. To that end, States Parties are encouraged to:

a. recognize communities, groups and individuals as the bearers of traditional knowledge about geoscience, particularly the climate;
b. foster scientific studies and research methodologies, including those conducted by the communities and groups themselves, aimed at understanding and demonstrating the effectiveness of knowledge of disaster risk reduction, disaster recovery, climate adaptation and climate change mitigation, that are recognized by communities, groups and, in some cases, individuals as part of their intangible cultural heritage, while enhancing the capacities of communities, groups and individuals to face challenges related to climate change that existing knowledge may not address;
c. adopt appropriate legal, technical, administrative and financial measures to:
    - promote access to and transmission of knowledge concerning the earth and the climate, that is recognized by communities, groups and, in some cases, individuals as part of their intangible cultural heritage, while respecting customary practices governing access to specific aspects of it;
    - fully integrate communities, groups and individuals who are bearers of such knowledge into systems and programmes of disaster risk reduction, disaster recovery and climate change adaptation and mitigation.

4.4 Best Practices for Climate Resilience

With the aim of publicizing the best practices with the possibility of replicating in urban environments, a selection of 11 methods used by ancestral cultures (Table 4.1) for the management of their necessary resources for survival in a sustainable manner without generating negative impacts on the environment was carried out. Environments they inhabited. To do this, it was analysed in terms of origin (country and region), start date, current status of the practice, original purpose (related to climate change or other), main challenges related to climate change and replicability possibilities, various sectors, type of intervention (hard technology, soft landscape-based, Nature-based solutions, Management practice).

Table 4.1 Best practices on traditional knowledge and intangible cultural heritage for climate change adaptation

No.	Practice/project name	Region	Scale	Who is doing the practice	Traditional practice start date	Timeline of specific project	Current state of practice	Sectors
1	Courtyard houses as passive cooling strategy	Iran, central area	National	Locals architects and engineers	6400–6000 BC	Present	Presently used/practices	Housing
2	Seaweed as a traditional construction material; the Vandkunste architects' house in Læsø, Denmark	Denmark	Local (Læsø island)	Vandkunsten architects	300–350 years ago	2012–2013	Presently used/practices	Housing
3	Clam gardens	Alaska, British Columbia and Washington State	Regional (Pacific Northwest)	Pacific Northwest Tribes (e.g. Swinomish Indian Tribal Community, Coast Salish tribes, Tribes in the Puget Sound region)	From late Holocene (11.700 years ago)	Present	Presently used/practices	Food

(continued)

Table 4.1 (continued)

No.	Practice/project name	Region	Scale	Who is doing the practice	Traditional practice start date	Timeline of specific project	Current state of practice	Sectors
4	Urban food gardens to adapt to climate changes	Malawi and Uganda	National	Manvester Ackson Khoza (National coordinator Malawi)/Slow Food Uganda with Youth Network		2019 - 2020	Presently used/practices and New practice with traditional food	Food
5	Tradition and science for the climate change adaptation indigenous climate knowledge in southern Uganda	Uganda	Region (South Uganda, Rakai District)	Farmers from the Rakai District	Thousands of years ago	2000s	Old practices adjusted to new CC	Food
6	Terraces agriculture as an adaptation to climate changes	Latino America and China	Regional				Presently used/practices	Food
7	Ngolo traditional cultivating practice in Tanzania	Tanzania	Regional—Southern part of Tanzania	Matengo people	1900s	–	Still practices	Agricultre, land management, water
8	Green infrastructure project in Peru with use of Mamanteo—indegenous canals to adress Lima's water shortage	Peru	City—Lima	Lima citizens	AD 600	March 2015	Partially abandoned, and clogged, partially used by pastures communities	Water, lands and resource management

(continued)

Table 4.1 (continued)

No.	Practice/project name	Region	Scale	Who is doing the practice	Traditional practice start date	Timeline of specific project	Current state of practice	Sectors
9	Qanat – underground water transport system	Iran and widespred to other coutries	National	Settlements,	Iran before 1000 BCE	–	Partially used by farmers and villages, partially abandoned and derelict	Water
10	Aflaj—traditional water distribution, irrigation system	Oman	National (villages and town)				Presently used/practices	Water
11	Floating gardens for	Bangladesh	Regional	Barisal, Gopalganj and Piropur district spearing to North–East and North–West of Bangladesh	100 years old	–	In practice	Water, land resource management; food

(continued)

Table 4.1 (continued)

No.	Methods/ measures	Subcategories/ subsectors	Practice description	Relation to climate	Main climate change related and other problems	Replicability possibilities	References
1	Technology	Traditional houses	Construction of traditional courtyard houses as a cooling strategy response to droughts and temperature changes	Practice established to response to climate challenges (warm)	Temperature increasing, droughts	Yes, in urban areas too, imitating criteria (orientation, dimensions and proportions, opaque and transparent surfaces)	Khalili and Amindeldar (2014), Asimakoupolos and Santamouris, 1996, Pérez-Lombard et al. (2008), Soflaei et al. (2016)
2	Technology	Building construction; construction material for modern buildings	Seaweed as traditional and low environmental impact construction material, used to climate change adaptation in modern buildings	Ancient practice established to response to climate challenges (cold)	Cold temperatures	Yes, in modern constructions (seaweed as a renewed and modern construction material that also respect old local traditions)	vandkunsten.com, Dahl (2008), Yang (2012), Widera (2014a, 2014b), Hansen (2013), renoteckroofing.com

(continued)

Table 4.1 (continued)

No.	Methods/ measures	Subcategories/ subsectors	Practice description	Relation to climate	Main climate change related and other problems	Replicability possibilities	References
3	Technology	Fishing local subsistence	Clam gardens delimitated by a wall that reduce the slope of the sea soil and create an ecosystem favorable to harvest clams. They provide a place for the tribe to continue their practice of harvesting clams for subsistence	Traditional practice adapted/ adjusted to respond to climate challenges (sea pollution)	Warming waters, increase of carbon in atmosphere	Yes	www.npr.org, www.yesmagazine.org, Gazeau et al. (2013), Groesbeck et al. (2014)
4	Community involvement / Measures and Management of droughts and floods	Crop production Food insecurity and community gardens	Creating sustainable and diversified indigenous food gardens Urban farming as a way to increase access to nutritious vegetables in homes	CC contributes to low performance in agriculture; During dry season they have to find new ways to cultivate food. Lack of water facilities for irrigation. Floods during the rainy season	Quality and diversification of agriculture products are affected by CC Droughts, high temperatures, floods during the rainy season	YES	www.slowfood.com

(continued)

Table 4.1 (continued)

No.	Methods/ measures	Subcategories/ subsectors	Practice description	Relation to climate	Main climate change related and other problems	Replicability possibilities	References
5	Management of agriculture	Crop production	This system of indigenous knowledge leads farmers to participate as agents as well as consumers in programs that use modern climate science to plan for and adapt to climate variability and climate change	Adapt traditional climate knowledge to climate variability and climate change	Precipitations, climate variability, temperatures	Yes, replicate the system anticipating variability also thought scientific instruments	Orlove et al. (2010), Green and Raygorodetsky (2010), Nyong et al. (2007a, 2007b)
6	Management of agriculture	Crop production	System of terrace to deal the climate change in sloping soils			Yes	Sun et al. (2012), Bocco and Napoletano (2017)
7	Nature-based solutions	Use of traditional practices in agricultrie	Cultivating in the steep slopes through pits dug in ridges; meant to conserve water and prevent soil erosion hence maintain the water and nutrient levels in the soil	Ancient practice for water and soil preservation	Soil erosion and loss of soil fertility; conserves water and nutritions thus counteracts the water deficit combating the effects of drought and land degradation	In villages, perhaps in cities of similar climate and proper inclination	Kato (2001), Allan (1965), Basehart, (1973), Kihila (2018), Vogeler et al. (2009), JICA (1998)

(continued)

Table 4.1 (continued)

No.	Methods/ measures	Subcategories/ subsectors	Practice description	Relation to climate	Main climate change related and other problems	Replicability possibilities	References
8	Nature-based solutions	Water management (storage and delay)	Indigenous infiltration enhancement system of canals diverting water from mountains to emerge in lower located springs and reservoir	Ancient practice in time of restoration along the wetland restoration	Water shortage for agricultural and domestic purposes and settlements, improve water security in mountain region	Can be combined with green infrastructure projects in Peru	Goodier (2019), Collyns (2015), Schatz (2015), Ochoa-Tocachi et al. (2019)
9	Hard infrastructure	Water and landscape management—water supply	Subterranean infrastructure—canals and shafts for accessing and managing underground water for farming, gardens, settlements	Ancient practice for accessing water for irrigation and urbanization	Water shortage in arid and semi-arid climate	New city, in other counties but similar environment, restoration of the existing	Ali Asghar Semsar Yazdi (2011), Beckers, et al. (2013), English (1968), Taghavi–Jeloudar and Han (2011), Taghavi-Jeloudar et al. (2013)

(continued)

Table 4.1 (continued)

No.	Methods/ measures	Subcategories/ subsectors	Practice description	Relation to climate	Main climate change related and other problems	Replicability possibilities	References
10	Technology + Management	Water distribution—water supply + agriculture	Canal system constructed above or below ground to collect under-ground water, water from natural springs, or water from the baseflow of wadi	Water provision first domestic use, and for irrigation in agricultu. Also industrial and other purposes, such as for water mills or to store water for travellers	Agriculture production (helps farmers to control drought and guarantees basic crop production), mcontrol and mitigation efforts against drought	Yes in similar climatic conditions	Remmington (2018)
11	Nature-based solutions	Agricultural production in floodplain areas	Floating gardening is an age-old practice of cultivation in the southern floodplains of Bangladesh; environment-friendly means to utilize the natural resources of wetlands to grow vegetables and other crops almost all-the-year-round	Useful in many wetlands during normal floods and also for rebuilding lives after disastrous floods through vegetable production, and by raising rice and vegetable seedlings to save time for subsequent winter cultivation	Extended water logging by improving food security	Yes, in other regions and countries of the water-logging conditions, and available local materials for constructions, and communities willing to adapt such method, proper climatic conditions for crop cultivation	Islam and Atkins (2007), Md Infranullah and Md Kalam Azad (2011), Biswas Chowdhury and Moore (2017), Winterborne (2005)

The purpose of this selection is to recognize the importance of the practices used by ancient cultures in managing climate change, and to publicize their ability to replicate and adapt to current environments. This recognition is an initiative to demonstrate the technologies and practices that current governments can use to deal with the problems generated by the climate emergency (floods, droughts, among others) (Table 4.2).

4.4.1 Courtyard: Traditional Building Solutions for Arid-Hot Climate Regions in Iran

The construction of courtyard's houses is a traditional and also actual practice, established to response to climate challenges. The first courtyard houses were built in the central Jordan Valley in ca. 6400–6000 BC. However, it has spread throughout the world, adapting to different contexts, such as Morocco, Spain, India and Japan. We will focus on the Iranian courtyard as an example of old practice that is presently used at national level, established to response to drought, strong winds and temperature changes from day to night.

Iranian architects and engineers have reasoned for many years on residential buildings adaptation to extreme climatic conditions. Iran, mostly in the central area, is characterized by arid-hot climate, with temperature ranging from 40 to 45 °C and 0 to 5 °C, low precipitations, humidity at 30% and dusty wind (Khalili & Amindeldar, 2014). Their main problem is the temperature change from day to night, which develops very different microclimates inside the houses during the day.

The solution for high amplitude is an efficient typology of house based on a central courtyard, with very long tradition. This structure allows to obtain a ventilated environment: the cool night air remain in the courtyard for the night, keeping the house fresh during the day and small courtyard are batter in storing cold air. It is a zero environmental impact technology, which does not consume energy, an important aspect because nowadays more than a third of all energy is spent on heating and cooling of buildings in any country (Pérez-Lombard et al., 2008).

Therefore, the traditional courtyard is an old passive cooling strategy to adapt to climate change that focuses on heat gain control and heat dissipation in a building in order to improve the indoor thermal comfort with low or nil energy consumption (Asimakoupolos and Santamouris, 1996). However, the advantage of courtyards in hot desert climates is to create a microclimate that takes advantage of wind flow and temperature change to ensure a cool and humid climate in the building that reduces energy consumption for the cooling; them also provides security and privacy for residents (Khalili & Amindeldar, 2014). This practice is not only about the shape, exposure and orientation of the building, but it regards technological advantages that can be gained from using technologies to tackle climate change.

Table 4.2 Graphic resources on best practices on traditional knowledge and Intangible cultural heritage for climate change adaptation

No.	Practice/project name	Images	Source and description
1	Courtyard houses as passive cooling strategy	A typical central courtyard in Rasoulian in Yazd  Thermal performance analysis of central courtyard at night and in day 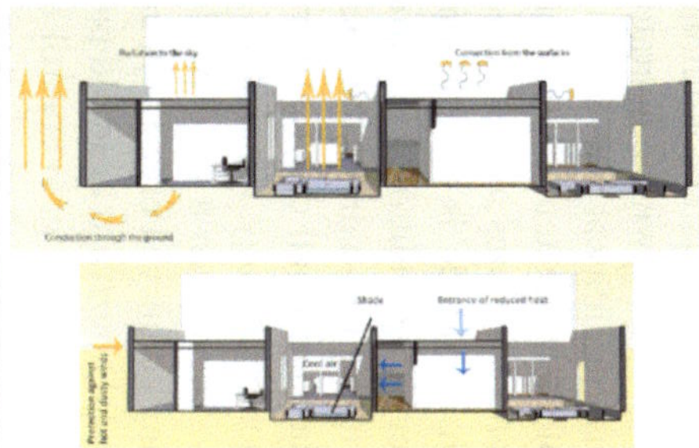	The ancient practices in the traditional architecture of Iran are described firstly by building patterns based on climatic considerations (such as cellars, patios, wind catchers, vaulted ceilings and porches) and secondly by behaviour patterns such as intelligent responses of great importance. Range to climate (i.e., behavioural adaptations) observed in past lifestyles. (Khalili & Amindeldar, 2014)

(continued)

Table 4.2 (continued)

No.	Practice/project name	Images	Source and description
2	Seaweed as a traditional construction material; the Vandkunste Architects' house in Læsø, Denmark	The Modern Seaweed House 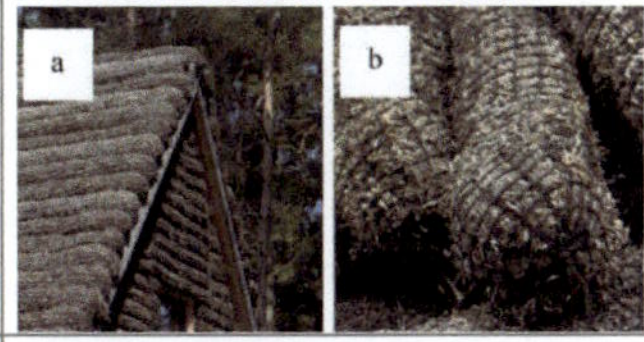	(a) The seaweed roof cladding and (b) the detail of seaweed placed in the knitted nets. The Modern Seaweed House (2012–2013) on Læsø, designed by Vandkunsten in cooperation with Realdania Byg. Photographer: Helene Hoeyer Mikkelsen/ Realdania Byg (July 2013)
3	Clam gardens	Clam garden images 	Clam garden images. (A) ancient clam gardens on Quadra Island, BC, Canada, are intertidal beach terraces built by humans by constructing (B) a rock wall at low tide typically between 0.7 and 1.3 m above chart datum. (C, D) Quadra Island clam gardens range in size and shape but generally create shallow sloping intertidal terraces encompassing tidal heights of 0.9–1.5 m above chart datum. https://doi.org/10.1371/journal.pone.0091235.g001 (Groesbeck et al., 2014)

(continued)

Table 4.2 (continued)

No.	Practice/project name	Images	Source and description
4	Urban food gardens to adapt to climate changes	Traditional practice in Uganda with the aim of mobilizing citizens to cultivate urban gardens 	Traditional practice in Uganda with the aim of mobilizing citizens to cultivate urban gardens as a way to increase access to nutritious and affordable fruits and vegetables among families and local communities

(continued)

Table 4.2 (continued)

No.	Practice/project name	Images	Source and description
7	Ngolo traditional cultivating practice in Tanzania	Intensive cultivation and environment use among the Matengo in Tanzania 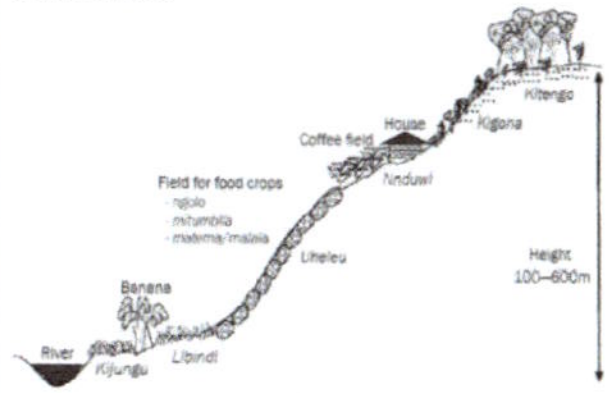 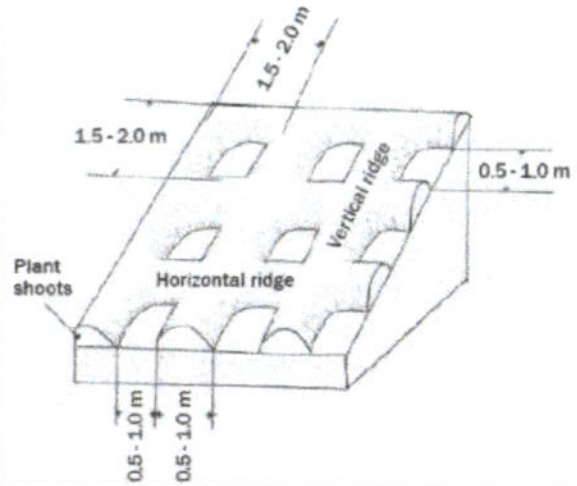	Kato (2001). Intensive cultivation and environment use among the Matengo in Tanzania. African Study Monographs, 22(2), 73–91
8	Mamanteo–prehispanic channels to address water shortage in Lima	Conceptual model of the pre-Inca infiltration enhancement system 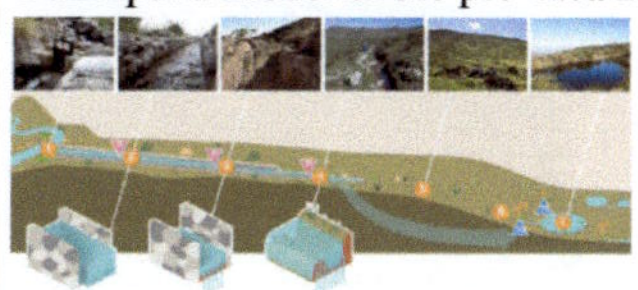	Ochoa-Tocachi et al. (2019) Potential contributions of pre-Inca infiltration infrastructure to Andean water security, Nature Sustainability 2, 584–593

(continued)

Table 4.2 (continued)

No.	Practice/project name	Images	Source and description
9	Qanat—traditional underground water supply system in Iranian plateau	Drawing of a hypothetical territorial section showing how the *qanat* system 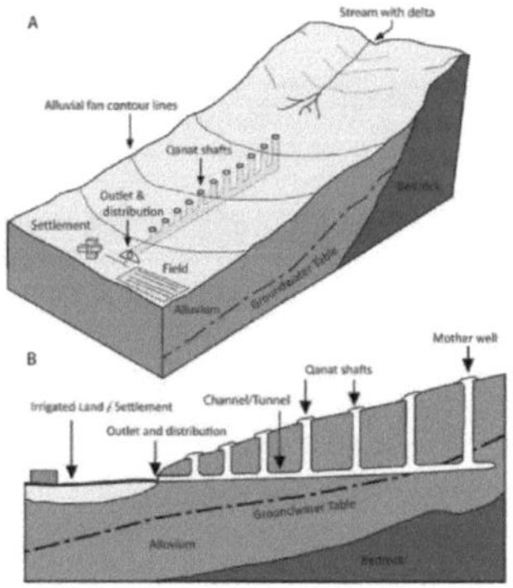	Drawing of a hypothetical territorial section showing how the *qanat* system and various architectural constructs such *bagh* (garden), *ab-anbars* (water storage), and *yakhchal* (ice storage) provided water for settlements and farms. It established a system that supported life in the territory Sanaan Bensi (2020) The Qanat System: A Reflection on the Heritage of the Extraction of Hidden Waters In Adaptive Strategies for Water Heritage: Past, Present and Future. Hein, C. (ed.). Cham, Springer, p. 41–57 17 p Beckers et al. (2013) Ancient water harvesting methods in the drylands of the mediterranean and western asia. Journal of Ancient Stududies. 2, 145–164

(continued)

Table 4.2 (continued)

No.	Practice/project name	Images	Source and description
10	Aflaj—traditional water distribution, irrigation system	Aflaj—traditional water distribution, irrigation system 	Using gravity, water is channelled from underground sources or springs to support agriculture and domestic use. The fair and effective management and sharing of water in villages and towns is still underpinned by mutual dependence and communal values and guided by astronomical observations. Numerous watchtowers built to defend the water systems form part of the site reflecting the historic dependence of communities on the aflaj system. Threatened by falling level of the underground water table, the aflaj represent an exceptionally well-preserved form of land use. https://whc.unesco.org/en/list/1207/
11	Floating gardens in Bangladesh	Floating gardens in Bangladesh 	Md Infranullah and Md Kalam Azad (2011) Floating gardens in Bangladesh: a, means to rebuild lives after devastating flood, Indian Journal of Traditional Knowledge 10 (1) 31–38. http://www.digital-development-debates.org/issue-06-innovation--science-and-technology--the-floating-gardens.html

This traditional solution of passive cooling can be replicated in low energy designs at present, in cities as well as in the countryside, where the climate is dry and very hot, aspects accelerated and extended by climate change.

4.4.2 *Seaweed as a Construction Material. An Old Building Tradition*

In 2012–2013 Vandkunsten Architects adopted traditional method of using seaweed as a construction material in a house. Its use has its origins in Læsø, a small island in Denmark, and it is an example of ancient practice, established 300–350 years ago, to response to climate challenges. The project name is "Modern Seaweed House" and it is a 90 m^2 summer house built for Realdania Byg Foundation.

Seaweed is an ancient material established to response to extreme temperatures in the absence of traditional building materials, such as wood (Yang 2012; Widera, 2014a, 2014b). As pointed by Torben Dahl "Throughout the world, the expressions of traditional architecture are based on and adapted to local conditions (Dahl, 2010). This applies primarily to the local availability of materials and the response to the climatic conditions" (Dahl, 2008, p.8). For instance, during the middle-ages the Læsø economy was based on salt, and a large amount of wood was required for its processing (Hansen, 2013), and the result was the loss of wood needed to build houses. The islanders' solution was a use of a seaweed called eelgrass, combined with driftwood, that lasts close to 300 years due to its high salt content which prevents decay and also acts as a fire retardant.

Today the structural system has evolved into knitted wool pillows stuffed with the seaweed that, other than being a sustainable material, as Søndermark (2013) noticed "reproduces itself every year in the sea, (…) comes ashore without any effort from humans, and it is dried on nearby fields by sun and wind". The solution is also non-toxic, with excellent insulating properties, CO_2 reducing and naturally regulates indoor humidity and fireproof.

Vandkunsten Architects highlights the use of seaweed as a material that respect both the old traditions and the environment. The house that they designed respect the local building traditions in size and orientation, the value of the isolation of the seaweed is close to mineral wool, it is in all walls and floor of the house and it was 'designed for disassembly'. The result is a house built with traditional materials, respecting the building traditions of the island, with high insulation and longevity.

Adaptation to climate change consists of the use of a durable and natural material with low environmental impact, that could be used as construction material for modern buildings in the countryside, as well as in cities, where this alga is present in nature.

4.4.3 Clam Gardens. Traditional Practice to Fight Climate Change

The tradition of building clam gardens was established in late Holocene in an international area along the coasts of Alaska, British Columbia and Washington State (Groesbeck et al., 2014). It is an old traditional practice from Pacific northwest Tribes, that was adjusted to respond to climate changes, mainly the sea pollution which stops the production of the seaweed on which the clams feed.

This practice regard clam harvest and local subsistence. Nowadays shellfish population face climate change challenges, as see level rise, increase of sea temperature, ocean acidification, and environmental pollution derived by use of plastic by commercial fishermen and bacterial pollution from upstream companies, with consequent effects on pH acidity levels. These criticalities are reflected in the abundance, growth rate and size of clams. Therefore, the rise in water temperature has reduced the production of algae on which clams feed.

As reported in the online magazine Northwest Treaty Tribes Magazine, the Swinomish tribe is reviving ancient clam gardens, fallen into disuse with colonization, to improve nearshore habitat and give tribal members more opportunities to harvest traditional food. They have built a clam garden in the intertidal area of the Swinomish Reserve to provide a place to continue the practice of clam harvesting for subsistence. A clam garden is a rock wall built so low down on a beach that it is only visible at low tide. This reef creates a buffer habitat that makes the water less acidic, allows the erosion of the beaches and the energy of the waves to be contained, creating an ecosystem suitable for clam reproduction.

The clam gardens could be relocated around the British Columbia, Alaskan, and Washington state coasts recovering abandoned sites or creating new ones in areas favourable to clam's growth. It is an opportunity to use ancient practices to ensure the livelihood of the community.

4.4.4 Tradition and Science for the Climate Change Adaptation. Indigenous Climate Knowledge in Southern Uganda

Farmers from the south Uganda use a system of indigenous knowledge combined with modern climate science to plan for and adapt to climate variability and climate change. It is an example of old practice, established thousands of years ago, now adjusted to deal new climate change, such as precipitations and climate variability, and that regards the management of agriculture activities.

Indigenous people are the main victims of climate change induced by industrialized societies, without being the main responsible (Green & Raygorodetsky, 2010). Farming communities use their indigenous traditions as a guide to address the future challenges of climate change, using a collective memory of drought and rainfall seasons (Nyong et al., 2007a, 2007b).

An example of these practices is Rakai District in southwestern Uganda (Orlove et al., 2010), characterized by a multi-ethnic population and consequently different indigenous traditions. Local farmers possess ancient knowledge regarding seasonal rainfall variability, understanding that timing and quantity of precipitation significantly impact their food supplies and income.

Nowadays, the traditional knowledge on climate is implemented by the Ugandan Department of Meteorology that is responsible for the data collection of the time series of weather events. Therefore, tradition and science work together to ensure the subsistence of the population and local communities.

The strength of Rakai District is the diversity of crops which enables it to diversify crops according to the challenges posed by climate variability and climate change. For instance, if the rainy season is late, part of cycle crops, as potatoes and maize, are planted in arid soils in anticipation of the rains; others as beans are cultivated on banana gardens as they both retain moisture for dry periods (Orlove et al., 2010).

Orlove et al. (2010) lists the indigenous skills and practices useful to the adaptation to seasonal rainfall variability, such as the familiarity with patterns of precipitation and temperatures, a set of local traditional climate indicators, the observation for meteorological events and the collection of information about season in others Uganda and Tanzania's regions. Therefore, the local communities use a collective memory pattern of timing and duration of the seasons to collect information regarding the amount of rain, clouds' shape and colour, wind and dates of seasons. In this way they should predict the weather forecast and adapt their cultivated fields.

Rakai's adaptation to the climate variability is an agricultural planning practice that includes a variety of spatial, temporal and social scales (Orlove et al., 2010) and local farmers showed a dynamic approach to climate variability and climate change. The replicability of the practice is the application of indigenous knowledge to direct scientific forecasting towards farmers' priorities.

4.4.5 Ngolo Traditional Cultivating Practice in Tanzania

The Ingulo or Ngolo system a traditional cultivating practice used by the farmers—Matengo people, that reside in the Mbinga District of the Ruvuma Region, in southern Tanzania. Ngolo has been practices in densely populated mountain villages since over 100 years. The Ngolo was invented originally by Mantego people to survive in harsh mountain region. Ngolo means "pit" in the Matengo language, and in literature it has been referred to "Matengo pit cultivation" (Kato, 2001).

Ingulo/ngolo is a practice used in agriculture, it is a combination of anti-erosion and soil fertility maintenance strategy of pits and ridges (Kihila, 2018). It is done by cultivating in the steep slopes through pits dug in ridges to grow the major food crops: maize and beans (Kato, 2001). The soil is affected in frequent torrential downpours during the year that have high rainfall, and the conventional tillage system causes a loss of soil and water and nutrient holding capacity and affecting soil fertility. Conversely the ngolo practice is geared towards conservation of water and avoiding soil erosion, thus maintaining the water and proper nutrient levels in the soil (Kato, 2001; Vogeler et al., 2009). The ngolo practice through preserving water counteracts the water deficit, it also improves soil moisture and water is utilized to sustain crop growth for a substantial time, enhances crop productivity and in general lead to sustainable water management (Kihila, 2018).

Ngolo fields are typically established on slopes of 5°–30°, and the average size of the field is about 0.7 ha, and the average size of a square grid ridge is about 2 m. which results in more than 1500 pits (Kato, 2001). Principally, a household would have two fields one for maize and one for beans, and the crops are cultivated in rotation (Kato, 2001).

Ngolo fields are cultivated in March, toward the end of the rainy season. It is done though first men cuts the weeds that have grown during the rainy months and, after a week, arrange the stalks in a square grid of 1.5–2.0 m. Further women dig up the holes within the limits of the grids and use soil to cover the stalks. The grid ridges produce many pits over an entire field. These pits preserve soil and water, while the covered weeds mature the soil (Allan, 1965; Basehart, 1973; JICA, 1998).

The strategy is targeting water and nutrient conservation; but is also effective strategy for combating the effects of drought and land degradation. The strategy is simple, flexible to implement; affordable and viable, and replicable, and makes it a sustainable strategy (Kihila, 2018).

4.4.6 Green Infrastructure Project in Peru with Use of Mamanteo—Indigenous Canals to Address Lima's Water Shortage

The Peruvian SUNASS, the national water regulator of Peru has started a programmed in 2015 for which $112 million of Lima's water fees was funneled to support the Andes mountain region in adapting to climate change and improve water quality. $22 out of the main fund is invested for the green infrastructure programmed, that include restoration and revival of the traditional infiltration's structures called Memento and restoration related highland wetlands. The green infrastructure project aims to address the challenge of water shortage in Limas in the dry season (Collyns, 2015; Hunter Burkett, 2019).

Mamanteo is an indigenous infiltration enhancement system to cope with climate variability, and bridge seasonal droughts in Peruvian Andes, to supplying irrigation water to agriculture and provide water supply to Lima's inhabitants (Collyns, 2015; Goodier, 2019; Schatz, 2015). Ochoca-Tocachi et al. (2019) describes Mamanteo as a *"nature-based water harvesting technologies" to manage drought risks under natural climatic extremes*. The canal system was developed by a civilization called the Wari, a pre-Incan empire culture, dating back to 1400 years. "Mamanteo" the Spanish word for "breastfeeding", or "amunas" in Quechua, an indigenous Andean language, translates to "retaining" (Ochoa-Tocachi et al., 2019, Hunter Burkett, 2019). The system is still used by some Andean peasant communities of Peru, however part of the system is clogged and abandoned.

The Mamanteo system consists of diversion canals, infiltration canals, infiltration hillslopes and springs and ponds, and can retain water from 2 weeks to 8 months, with an average of 45 days.

The operation of the system starts when the irrigation canals system captures and directs hydrological runoff/river water during the wet season, from December to April, from headwaters streams onto mountain slopes where it infiltrates. Consequently, water percolates through cracks and aquifers to then emerge in springs and natural reservoirs/ponds during the dry season when water supplies are scarce, from August till October (Hunter Burkett, 2019). Water delayed by extended subsurface retention time cause the increased yield and reliability of downslope springs during the dry months (Ochoca-Tocachi et al., 2019).

Lima is affected by the floods due to heavy rains during the wet season and is the world's second-largest desert city in the dry season causing water shortage. Research shows that the restoration and replication of these systems in line with to the water demand Lima, could ensure diverting and delaying 35% of the water during wet season. The system will need to be part of a more comprehensive water management strategy in the rural region and can be coupled with the existing grey infrastructure in Lima to increase water security (Ochoca-Tocachi et al., 2019).

The implementation of nature-based solutions that integrates traditional and indigenous knowledge and practice can address the challenge of water supply to large populations in hydrologically variable and arid environments and improve water security and climate change adaptation in mountain regions. The implementation of solution can also help in achieving the sustainable development goals. (Ochoa-Tocachi et al., 2019).

4.4.7 Qanat—Traditional Underground Water Supply System in Iranian Plateau

Qanat is a technical solution of tunnels for harvest and managing groundwater in arid and semi-arid areas to tackle the challenge of scare water resources, dating back to 3000 years. It is believed the qanat system was invented in Iran (former Persia), where currently there is about 22,000 qanats (Sanaan Bensi, 2020). This ancient technology was extensively widespread and exists in more than 35 countries known as *Karez* in Afghanistan, *Galeria* in Spain, *Khotara/Khetarra* in Morocco, *Aflaj* in Oman, *Foggara* in North Africa, *Kanerjing* in China, and *Auon* in Saudi Arabia and Egypt (Taghavi-Jeloudar et al., 2011, Beckers et al., 2013).

Qanats it is subterranean infrastructure that gave access to the hidden water at the foot of mountains. It consists of vertical shafts of gradually increasing depth connected by a horizontal underground conduits or tunnel (dehliz) beneath earth surface, with a purpose to convey the water from subterranean sources down a slight slope to the earth surface by using gravity flow conditions. Qanats address the challenge of water shortage for irrigation of farms, domestic use and settlements, and it has shaped the landscape, organized territories. Consequently, was the foundation of inhabitation, construction, and prosperity in Iran (Sanaan Bensi, 2020; Taghavi-Jeloudar et al., 2011). The law of *harim or* borders was protecting the right of ownership; and defined frames for protecting underground water resources, by forbidding the building of new motherwells or any other construction within a defined distance of an existing qanat (English, 1968; Sanaan Bensi, 2020).

Since qanat rely on the gravity, it does not extract water from less elevated spots of aquifer, and thus does not deplete the groundwater resources of aquifers (Ali Asghar Semsar Yazdi, 2011). Some old qanats are still in use today in Iran, primarily in villages and smaller towns. However, parts of these systems have been abandoned. The efficiency of qanats was impacted when pumping wells were introduced during the Second World War (Taghavi-Jeloudar et al., 2011) and caused over-exploitation of groundwater decreasing the water table. (Ali Asghar Semsar Yazdi, 2011).

Qanat has a potential for replication to benefit the arid and semi-arid regions without sufficient permanent river flow. The existing the existing qanats require restoration, which can be complemented with the use of new materials and modern geological techniques (Taghavi–Jeloudar and Han, 2011).

4.4.8 Floating Gardens in Bangladesh

Floating gardens is an indigenous agricultural practice that has been originally cultivated in southern floodplain of Bangaldesh (Barisal, Gopalganj and Piropur district) for over 100 years. It is an adaptation strategy that support communities to sustain lives and livelihoods under flooded and long-term waterlogged conditions. Over the last decade the practices have been replicated and scaled up in other flood-plain regions of Bangladesh in the North–East and North–West for nutritional security with support on non-governmental organizations (Md Infranullah & Md Kalam Azad, 2011). In Bangladeshi language, the method is traditionally called *dhap* (Islam & Atkins, 2007).

Floating agriculture is a technique of cultivating crops and vegetables on a soilless floating platform or raft made of construction materials: decomposing heaps of water hyacinth and other aquatic weeds. The scientific term for it is 'hydroponics,' which is a soilless cultivation technique of growing plants and raising seedlings by using *inert growing medium, such as gravel, sand, perlite, vermiculite, clay pebbles,* etc., *where plants receive essential nutrients for their growth either from a nutrient solution or organic materials added to the medium* (Biswas Chowdhury & Moore, 2017; Winterborne, 2005).

The size of the platforms varies, construction of 1 m^2 floating platform requires 5 m^2 of water. The cultivation of floating gardens with monsoon crops takes place from June–July until end of September–October. With the flood water recession in November–December, the platforms are dismantled, and the residuals are used for preparing beds on the soul for winter crops cultivation. Since, the water hyacinth is an invasive species, abundant in wetland of Bangladesh, this is an environment-friendly method using locally available natural recourses to cultivate plans and crops almost all year round (Biswas Chowdhury & Moore, 2017; Md Infranullah & Md Kalam Azad, 2011), as well as economically affordable, and socially viable. It provides benefits in terms of food security, land use capacity, health, and household income.

The technology is not suitable to all open water bodies, those with frequential tidal interruptions, it is fragile in terms of devastating floods and strong waves (Biswas Chowdhury & Moore, 2017; Md Infranullah & Md Kalam Azad, 2011). However, it can be supporting for rebuilding lives and ensuring food security after the disastrous flood with extended water logging, and for regular floods.

The technology has also a potential as a community-based adaptation strategy in Bangladesh to climate change (Md Infranullah & Md Kalam Azad, 2011). This technology can be replicated and used in countries with similar water-logging conditions, and available local materials for constructions, and communities willing to adapt such method, proper climatic conditions for crop cultivation (Biswas Chowdhury & Moore, 2017).

4.5 Conclusions

This review highlights the ancestral practices in response to the climate emergency that we are currently experiencing, revealing real practices that indigenous peoples have used and use today to adapt to the climatic difficulties presented by the areas where they are located. With the aim of improving the resilience and adaptation of cities, it is imperative to raise awareness and establish financing funds for the study and implementation of pilots based on ancestral practices.

The practices that ancestral communities possess and use to prepare for, mitigate, respond to, and recover from the impacts caused by climate change, serve as the optimal solution to the problems that contemporary peoples are going through. From hazard forecasts to anticipatory measures that are determined by social cohesion and belief systems.

Living within ecosystems offers an in-depth comprehension of them, underscoring the importance of actively sharing research outcomes in conferences, congresses, and analogous settings focused on mitigating, adapting to, and fortifying resilience against climate change. It's crucial to highlight the immense value in learning from the ancestral wisdom and exceptional biodiversity management practices of indigenous communities.

References

Aguilera-Díaz, M. M. (2004) La Mojana : riqueza natural y potencial económico. Doc. Trab. Sobre Econ. Reg. y Urbana *48* (2004) [Online]. Available: http://repositorio.banrep.gov.co/handle/20.500.12134/3204

Ali Asghar S. Y. (2011). Qanat; an ancient technique for adapting to new climate changes. *Journal of Arid Land Studies, 21*(2), 71–73. ICAL 1/DT X Joint International Symposium with JAALS.

Allan, W. (1965). *The African husbandman.* Edinburgh: Oliver & Boyd.

Architectural Seaweed. (n.d.). Retrieved August 14, 2022, from https://vandkunsten.com/en/projects/seaweedhouse

Arias, M. E., & Brown, M. T. (2009). Feasibility of using constructed treatment wetlands for municipal wastewater treatment in the Bogotá Savannah, Colombia. *Ecological Engineering, 35*(7), 1070–1078. https://doi.org/10.1016/J.ECOLENG.2009.03.017

Asimakopoulos, D., & Santamouris, M. (1996). Passive cooling of buildings. Earthscan. ISBN: 1873936478, 9781873936474.

Bardsley, D. K., & Wiseman, N. D. (2016). Socio-ecological lessons for the anthropocene: Learning from the remote indigenous communities of Central Australia. *Anthropocene, 14*, 58–70. https://doi.org/10.1016/j.ancene.2016.04.001

Bardsley. D. K. (2017). *Indigenous knowledge and practice for climate change adaptation* (Vol. 1–5). Elsevier Inc.

Basehart, H. W. (1973). Cultivation intensity, settlement patterns, and homestead forms among the Matengo of Tanzania. *Ethnology, 12*, 57–73

Beckers, B., Berking, J., & Schutt, B. (2013). Ancient water harvesting methods in the drylands of the mediterranean and western Asia. *Jurban of Ancient Stududies., 2*, 145–164.

Berkes, F. (1999). *Sacred ecology: Traditiona/ecologica/knowledge and resource management.* Philadelphia, PA: Taylor & Francis.

Berkes, F. (2009). Indigenous ways of knowing and the study of environmental change. *Journal of the Royal Society of New Zealand, 39*(4), 151–156. https://doi.org/10.1080/03014220909510568

Biswas Chowdhury, R., & Moore, G. A. (2017). Floating agriculture: A potential cleaner production technique for climate change adaptation and sustainable community development in Bangladesh. *Journal of Cleaner Production, 150*, 371–389.

Bocco, G., & Napoletano, B. M. (2017). The prospects of terrace agriculture as an adaptation to climate change in Latin America. *Geography Compass, 11*(10), Article e12330. https://doi.org/10.1111/GEC3.12330

Collyns, D. (2015). *Peru harnesses ancient canal system to tackle Lima water shortage.* The Guardian, entered: November 2020

Cutter, S. L., Mitchell, J. T., & Scott. M. S. (2000). Revealing the vulnerability of people and places: a case study of Georgetown County, South Carolina. *Annals of the Association of American Geographers, 90*(4), 713–737.

Dahl, T. (2008). *Climate and architecture.* London, New York: Taylor & Francis.

Dahl, T. (2010). *Climate and architecture.* Routledge. https://www.routledge.com/Climate-and-Architecture/Dahl/p/book/9780415563093

Darwin, C. (1859). *On the origin of species by means of natural selection, or, the preservation of favoured races in the struggle for life.* London: J. Murray; WWF

Davidson, K., Briggs, J., Nolan, E., Bush, J., Håkansson, I., Moloney, S. (2020). The making of a climate emergency response: Examining the attributes of climate emergency plans. *Urban Climate, 33*(January), 100666. https://doi.org/10.1016/j.uclim.2020.100666

Dowie, M. (2009). *Conservation refugees: The hundred-year conflict between global conservatíon and native peoples.* Cambridge, Mass: MIT Press.

Edible Cities: Rebuilding Food Sovereignty one Rooftop at the Time—Slow Food International. (n.d.). Retrieved August 14, 2022, from https://www.slowfood.com/edible-cities-rebuilding-food-sovereignty-one-rooftop-at-the-time/

English, P. W. (1968). The origin and spread of qanats in the old world. *Proceedings of the American Philosophical Society, 112*(3), 170–181.

FAO. (2019). SISTEMA AGROECOLÓGICO TRADICIONAL SISTEMA HIDRÁULICO DE AGRICULTURA ANFIBIA ZENÚ SEMBRANDO CAPACIDADES COOPERACIÓN BRASIL-COLOMBIA-FAO.

Gallardo, M. C. (2016). Socio-ecological inequality and water crisis: Views of indigenous communities in the alto Loa area. *Environmental Justice, 9*(1), 9–14. https://doi.org/10.1089/env.2015.0023

Gazeau, F., Parker, L. M., Comeau, S., Gattuso, J. P., O'Connor, W. A., Martin, S., Pörtner, H. O., & Ross, P. M. (2013). Impacts of ocean acidification on marine shelled molluscs. *Marine Biology, 160*(8), 2207–2245. https://doi.org/10.1007/S00227-013-2219-3

Goodier, R. (2019). Ancient solutions to weather seasonal drought. *Mechanical Engineering, 141*(11), 20–21.

Green, D., & Raygorodetsky, G. (2010). Indigenous knowledge of a changing climate. *Climatic Change, 100*(2), 239–242. https://doi.org/10.1007/S10584-010-9804-Y

Groesbeck, A. S., Rowell, K., Lepofsky, D., & Salomon, A. K. (2014). Ancient clam gardens increased shellfish production: Adaptive strategies from the past can inform food security today. *PLoS ONE, 9*(3), Article e91235. https://doi.org/10.1371/JOURNAL.PONE.0091235

Hansen, J. M. (2013). The salt industry on the Danish Kattegat island of Læsø (1150–1652): Hypersaline source, climatic dependence, and environmental impact. *Journal of Geography, 110*(1), 1–24. https://doi.org/10.1080/00167223.2010.10669493

Harnessing Indigenous Crops: Adapting to Climate Change with Traditional Knowledge—Slow Food International. (n.d.). Retrieved August 14, 2022, from https://www.slowfood.com/harnessing-indigenous-crops-adapting-to-climate-change-with-traditional-knowledge/

Hulme, M. (2009). *Why we disagree about climate change: Understanding controversy, inaction and opportunity* (p. 2009). New York: Cambridge University Press.

Hunter Burkett, M. (2019). Silent and unseen: Stewardship of water infrastructural heritage. In C. Hein (Ed.) *Adaptive strategies for water heritage: Past, present and future* (pp. 41–57). Springer: Cham

Islam, T., & Atkins, P. (2007). Indigenous floating cultivation: A sustainable agricultural practice in the wetlands of Bangladesh. *Development in Practice, 17*, 1.

Jackson, R. C., Dugmore, A. J., & Riede, F. (2018). Rediscovering lessons of adaptation from the past. *Global Environmental Change, 52*(May), 58–65. https://doi.org/10.1016/j.gloenvcha.2018.05.006

JICA. (1998). *Integrated agro-ecological research of the Miomb woodlands in Tanzania*. Final Report. Japan International Cooperation Agency, Tokyo.

Kato, M. (2001). Intensive cultivation and environment use among the Matengo in Tanzania. *African Study Monographs, 22*(2), 73–91.

Kettlewell, R. W. (1965). The African husbandman by W. Allan Edinburgh: Oliver & Boyd (1965), 63s. *Experimental Agriculture, 1*(4), 320–320. https://doi.org/10.1017/S0014479700021621

Khalili, M., & Amindeldar, S. (2014). Traditional solutions in low energy buildings of hot-arid regions of Iran. *Sustainable Cities and Society, 13*, 171–181. https://doi.org/10.1016/j.scs.2014.05.008

Kihila, M. J. (2018). Indigenous coping and adaptation strategies to climate change of local communities in Tanzania: A review. *Climate and Development, 10*(5), 406–416.

Leonard, S., Parsons, M., Olawsky, K., & Kofod, F. (2013). The role of culture and traditional knowledge in climate change adaptation: Insights from East Kimberley, Australia. *Global Environmental Change, 23*(3), 623–632. https://doi.org/10.1016/j.gloenvcha.2013.02.012

Liquete, C., Cid, N., Lanzanova, D., Grizzetti, B., & Reynaud, A. (2016). Perspectives on the link between ecosystem services and biodiversity: The assessment of the nursery function. *Ecological Indicators, 63*, 249–257. https://doi.org/10.1016/J.ECOLIND.2015.11.058

Margulis, L. (1998). *Symbiotic planet: A new look at evolution*. NewYork: Basic Books.

Md Infranullah, H., Md Kalam Azad, A. (2011). Floating gardens in Bangladesh: A means to rebuild lives after devastating flood. *Indian Journal of Traditional Knowledge, 10*(1), 31–38.

Megdiche-Kharrat, F., Ragala, R., Moussa, M. (2019). Promoting a sustainable traditional technique of aquifer water acquisition common to arid lands: A case study of Ghassem Abad Qanat in Yazd Province (Iran). *Water Science & Technology, 19*(2), 527–535.

Mohamed Shaffril, H. A., Ahmad, N., Samsuddin, S. F., Samah, A. A., Hamdan, M. E. (2020). Systematic literature review on adaptation towards climate change impacts among indigenous people in the Asia Pacific regions. *Journal of Cleaner Production, 258*, 120595. https://doi.org/10.1016/j.jclepro.2020.120595

Nyong, A., Adesina, F., & Osman Elasha, B. (2007). The value of indigenous knowledge in climate change mitigation and adaptation strategies in the African Sahel. *Mitigation and Adaptation Strategies for Global Change*. https://doi.org/10.1007/s11027-007-9099-0

Nyong, A., Adesina, F., & Osman Elasha, B. (2007). The value of indigenous knowledge in climate change mitigation and adaptation strategies in the African Sahel. *Mitigation and Adaptation Strategies for Global Change, 12*(5), 787–797. https://doi.org/10.1007/S11027-007-9099-0

Ochoa-Tocachi, B. F., Bardales, J. D., Antiporta, J., Pérez, K., Acosta, L., Mao, F., Zulkafli, Z., Gil-Ríos, J., Angulo, O., Grainger, S., Gammie, G., De Bièvre, B., & Buytaert, W. (2019). Potential contributions of pre-Inca infiltration infrastructure to Andean water security. *Nature Sustainability, 2*, 584–593.

Orlove, B., et al. (2010). Indigenous climate knowledge in southern Uganda: The multiple components of a dynamic regional system. *Climatic Change*. https://doi.org/10.1007/s10584-009-9586-2

Orlove, B., Roncoli, C., Kabugo, M., & Majugu, A. (2009). Indigenous climate knowledge in southern Uganda: the multiple components of a dynamic regional system. *Climatic Change, 100*(2), 243–265. https://doi.org/10.1007/S10584-009-9586-2

Pacific Northwest Tribes Face Climate Change with Agricultural Ancient Practice : NPR. (n.d.). Retrieved August 14, 2022, from https://www.npr.org/2019/10/08/767896285/pacific-northwest-tribes-face-climate-change-with-agricultural-ancient-practice?t=1603199718965&t=1603624411574

Pérez-Lombard, L., Ortiz, J., & Pout, C. (2008). A review on buildings energy consumption information. *Energy and Buildings, 40*(3), 394–398. https://doi.org/10.1016/J.ENBUILD.2007.03.007

Raygorodetsky, G. (2018). *Indigenous peoples defend Earth's biodiversity-but they're in danger* [online]. National Geographic. Accessed March 25, 2019. https://www.nationalgeographic.com/environment/2018/11/can-indigenous-land-stewardship-protect-biodiversity-/

Remmington (2018) Transforming tradition: The aflaj and changing role of traditional knowledge systems for collective water management. *Journal of Arid Environments, 151*(2018), 134–140

Sanaan Bensi, N. (2020) The qanat system: A reflection on the heritage of the extraction of hidden waters. In C. Hein (Ed.) *Adaptive strategies for water heritage: Past, present and future*. (pp. 41–57). Springer: Cham

Santamouris, M. (Matheos), & Asimakopoulos, D. N. (Demosthenes N.). (1996). *Passive cooling of buildings* (p. 472).

Schatz, J. (2015). *Lima taps ancient Andean canals to help fight its water shortage*. PRI, entered: November 2020

Shaffril, H. A. M., Ahmad, N., Samsuddin, S. F., Samah, A. A., & Hamdan, M. E. (2020). Systematic literature review on adaptation towards climate change impacts among indigenous people in the Asia Pacific regions. *Journal of Cleaner Production, 258*, Article 120595.

Soflaei, F., Shokouhian, M., & Mofidi Shemirani, S. M. (2016). Investigation of Iranian traditional courtyard as passive cooling strategy (a field study on BS climate). *International Journal of Sustainable Built Environment, 5*(1), 99–113. https://doi.org/10.1016/J.IJSBE.2015.12.001

Søndermark, J. (2013). In Frearson, A. *The modern seaweed house by Vandkunsten and Realdania Byg*. Retrieved from Dezeen Magazine, published July 10th, 2013. Retrieved October 10, 2020, from http://www.dezeen.com/

Sun, Y., Zhou, H., Zhang, L., Min, Q., & Yin, W. (2012). Adapting to droughts in Yuanyang Terrace of SW China: insight from disaster risk reduction. *Mitigation and Adaptation Strategies for Global Change, 18*(6), 759–771. https://doi.org/10.1007/S11027-012-9386-2

Swinomish Revives Ancient Clam Gardens to Fight Climate Change | Northwest Treaty Tribes. (n.d.). Retrieved August 14, 2022, from https://nwtreatytribes.org/swinomish-revives-ancient-clam-gardens-to-fight-climate-change/

Taghavi–Jeloudar, M., Han, M. (2011). Qanat, the masterpiece for climate change adaptation in arid area. In *Conference: The 1st international conference on green environmental technology.* KSEE, Incheon, S. KOREA

Taghavi-Jeloudar, M., Han, M., Davoudi, M., & Kim, M. (2013). Review of ancient wisdom of qanat, and suggestions for future water management. *Journal of Environmental Engineering and Science, 18*(2), 1–7.

The Modern Seaweed House by Vandkunsten and Realdania. (n.d.). Retrieved August 14, 2022, from https://www.dezeen.com/2013/07/10/the-modern-seaweed-house-by-vandkunsten-and-realdania/

Vogeler, I., Rogasik, J., Funder, U., Panten, K., & Schnug, E. (2009). Effect of tillage systems and P-fertilization on soil physical and chemical properties, crop yield and nutrient uptake. *Soil and Tillage Research, 103*(1), 137–143

Watson, J. (2019). *Lo-TEK design by radical indigenism.* TASCHEN ed.

Why Coastal Tribes Are Growing Clam Gardens That Look Like Asian Rice Fields—YES! Magazine. (n.d.). Retrieved August 14, 2022, from https://www.yesmagazine.org/environment/2017/07/06/coastal-tribes-garden-clams-to-preserve-culture-improve-health

Widera, B. (2014a). Possible application of seaweed as building material in the modern seaweed house on Læsø. In *30th international PLEA conference: Sustainable habitat for developing societies: Choosing the way forward—Proceedings.*

Widera, B. (2014b). *Possible application of seaweed as building material in the modern seaweed house on Laesø.* https://doi.org/10.13140/RG.2.1.1638.2881

Winterborne, J. (2005). *Hydroponics: Indoor Horticulture.* Surrey: Pukka Press.

Yang, J. (2012). Properties analysis of seaweed as a traditional building material. *Advanced Materials Research, 450–451*, 154–157. https://doi.org/10.4028/SCIENTIFIC5/AMR.450-451.154

Chapter 5
Shaaree na Reujade Panapata Akune Juapa [Living the Force of our Territory with the Mighty Power—Vivimos la Fuerza de Nuestro Territorio Con el Poder Superior]: Adaptation and Resilience in the *Eperara Siapidara* of the Colombian Pacific Coast

Jacinto Mocho, Jairo Tocancipá-Falla, and Romilio Tovar

Abstract In this chapter we examine the interrelationships which exist between situated knowledge and resilience-resistance in the *Eperara Siapidara* of the Colombian Pacific coast. We argue that these concepts are to be understood in their dynamics rather than in their particular static condition; that is, as processes and not as closed concepts. Methodologically speaking, the text is the result of several meetings of reading and joint interpretation that depart from the experiences of the Eperara Siapidara and partially from the university academic world. In principle, we consider it to be an intercultural and intertextuality exercise, if we assume that social reality can be understood as a form of text. In the early stages of discussions, it became clear that our perspective echoed Clifford Geertz's idea of an "ethnography of thinking" that deals not only with exalting "diversity but to take it seriously as itself an object of analytic description and interpretative reflection" (Geertz, 1983 2000:154). We

This chapter is the result of several work sessions by the authors where the events reported by Romilio and Jacinto were addressed, while Jairo questioned the narratives and stories told by them and wrote the paper, based on the genealogy of terms such as resilience, resistance and heritage, while the text was taking its own course. We would like to thanks to Rebecca Fullerton for her time in reviewing the text. Our interpretations do not commit her dedicated effort.

J. Mocho · R. Tovar
Nueva Bella Vista Indigenous Community, Eperara Siapidara, Municipality of Guapi, Cauca, Colombia
e-mail: jacintomocho@gmail.com

R. Tovar
e-mail: mona-ale96@hotmail.com

J. Tocancipá-Falla (✉)
Comparative Research Group of the University of Cauca, Cauca, Colombia
e-mail: jtocancipa@unicauca.edu.co

J. Morato et al. (eds.), *Intangible Cultural Heritage and Traditional Knowledge Related to Urban Resilience*, Resilient Cities, https://doi.org/10.1007/978-3-032-12713-6_5

also integrated the idea of co-theorization postulated by Professor Rappaport in her work with the Nasa in Colombia (2005a), which in our case, although not horizontal due to the nature of knowledge in the interaction, proposes an interrelation dissimilar and complementary in knowledge. Our aim is thus as follows: to unite the pieces of a puzzle where three categories, resilience, resistance and heritage, converge in a set of narratives associated with the history of the Eperara Siapidara. Finally, we do not conceive of the task of this dialogue as closed: rather, it opens new possibilities and conditions to refine ideas that, over time, can be assessed and adjusted. In other words, the ideas expressed here are also subjected to a resilience-resistance test.

Keywords Resilience · Resistance · Adaptation · Narrative · Indigenous knowledge

5.1 Introduction to Resilience, Resistance and Heritage: Sketches of the Same Path

This work represents an intercultural exercise in which academia and immaterial heritage converge with concepts of resilience and resistance. The dialogue of this paper poses as an invitation to explore the interactions and cultural expressions where these concepts meet, but also where they cross boundaries of knowledge and comprehension. We believe that, although these concepts are complex and difficult to comprehend, connecting and relating them to the social life of people and indigenous groups like the *Eperara Siapidara* can open new vistas. In raising these questions, we found ourselves on a common ground of enquiry.

For the purpose of this chapter, the triad resilience, resistance and heritage, posits questions and broad claims that warrant contextualization. Each of these concepts propose approaches, divergences and convergences that request examination. Here we present a general outline, in the understanding that a deeper and more detailed work is still required to complement and diversify the approaches made within it.

Let us begin with the first expression, resilience, which requires approaching several domains of knowledge. Its roots and trajectory in the Spanish domain show that it was originally conceived in the field of engineering and industry associated with the strength of materials: "steel with high resilience and wear resistance, and non-deformable" (Española 2018: s.f.). Although the expression resilience comes from the Latin *resiliens*, of *resilīre* which means "to jump back", "bouncing", "fall back", "to go back jumping" (Fraile, 1984; Española, 2001:s.f.), its use became more frequent in the first half of the twentieth century in the industrial and manufacturing fields.

The use of the term resilience in the social sciences, some authors contend, came relatively late in comparison to the field of engineering from which it arose. In this regard, Kotliarenco, Cáceres and Fontecilla point out:

In the late 1970s [1970], conversations began in a new domain, related to the development within the social sciences of the concept of resilience. The discussion around this concept began in the field of psychopathology, a domain in which it was found with great amazement and interest that some of the children raised in families in which one or both parents were alcoholics, and that during the development process of their children, they did not present deficiencies in the biological or psychosocial field, but on the contrary, they reached an 'adequate' quality of life (Werner, 1989). (quoted in Kotliarenco, Cáceres et al., 1997:1).

Since then, its use has become common in different areas that extend not only to individuals, but also to collectives and communities as a quality of elastic, adaptive movement that develops in the face of hostile or oppressive conditions. In a more comprehensive context, the *Diccionario de la Real Academia de la Lengua Española* defines resilience in two ways:

1. The ability of a living being to adapt to a disturbing agent or an adverse state or situation.
2. The ability of a material, mechanism or system to recover its initial state when the disturbance to which it had been subjected has ceased.

The Oxford English Dictionary (2020) offers a similar definition:

1. The ability of people or things to recover quickly after something unpleasant, such as shock, injury, etc.
2. The ability of a substance to return to its original shape after it has been bent, stretched or pressed. (https://www.oxfordlearnersdictionaries.com/us/definition/english/resilience?q=resilience, reviewed on the eighth of march 2020)

In the Spanish version, the second concept of resilience is associated with "things", i.e. the materials industry, while in the English version, the first definition incorporates both "people" and "things". However, the second definition in English is shared with the Spanish, focused solely on "things". Similarly, in other languages, such as German, resilience also adopts the generic Latin form - "*psychische Widerstandskraft; Fähigkeit, schwierige Lebenssituationen ohne anhaltende Beeinträchtigung zu überstehen*: "the psychological strength to resist, to overcome difficult situations in life without permanent harm" (https://www.duden.de/rechtschreibung/Resilienz, reviewed on the 28th of November 2019, we thank Professor Josef Drexler for the translation). In this case, the addition of the prefix "psycho" implies a living entity that is facing a stressful condition. Resilience then emerged in the physical domain of "things" which, despite pressure of any kind, are able to return to their original shape and form. A preliminary reference is that its basic use in the domain of physics seems to be transferred to the domain of human beings, or the social field.

The second term, resistance, also presents a polysemy field that has been approached from the same philosophy to indicate how the subject, with his will and impulse, is built in opposition or resistance to the world and physical bodies (Mora, 1994).

In the last few decades, however, the expression 'resistance' has been focused on actors, capacities, processes or phenomena that oppose being naturalized, controlled, dominated or domesticated by other local, regional, national or global actors, capacities or processes (e.g. capitalist regime, private capital or interests of groups that control the state, etc.).

The *Corpus Nuevo Diccionario del Español* (online version 2019) indicates that this term appeared in the fourteenth century in association with wars and war confrontations where the dominant actor subject prevails.

This definition grants nine values of meaning, ranging from generalities such as a field, capacity and group of people who resist or oppose an invader, to that of psychoanalysis to indicate how the patient opposes their unconscious impulses. Other definitions are associated with subjects such as electricity, physics and mechanics, where it is understood as an element of opposition in a circuit, a magnitude or force that is in relation to another force or physical interference (Española, 2001). Stemming from Latin, the term resistance bears different meanings which share a common genealogy: "*rĕsisto, ĭs, sistĕre, restĭti.* (from *re* y *sisto*). v. intr. To stop, hold back, go no further, stop the pace, stop (…) // Cic. stand firm, keep self-control. / / Face, resist, oppose, resist" (Fraile, 1984:425). In the Oxford English Dictionary, the word resistance appears with six notions, most of them associated to "dislike of or opposition to a plan, an idea, etc.; the act of refusing to obey; the act of using force to oppose somebody/something; **resistance (to something)** the power not to be affected by something; **resistance (to something)** a force that stops something moving or makes it move more slowly; the opposition of a substance or device to the flow of an electrical current; a secret organization that resists the authorities, especially in a country that an enemy has control of" (https://www.oxfordlearnersdictionaries.com/us/definition/english/resilience?q=resilience, reviewed eighth March 2020). Similarly, to resilience, the first set of notions is associated with the domain of human beings who react to a pressure by resisting it, however not referring to the condition of the subject once the pressure ceases, as in the case of resilience.

However, resistance does not always imply confrontation, which is why 'passive resistance' is also defined as "the reluctance to do or accomplish something" (Española, 2001: 1328). More recently, in the context of social, student and indigenous movements, resistance has become an active and passive word that constitutes part of the discourse and struggles against the State, institutions and multinationals.

These definitions in different languages help us to understand resilience as a dual concept: although in its initial state, the object or subject maintains a firmness or other certain characteristic, under pressure or disturbance, that object or subject becomes "elastic" or malleable before returning to its initial state. This does not concern whether, under a certain pressure, the object or subject changes its configuration, but rather the capacity not to succumb to the disturbing pressure that could either destroy the object or de-structure the subject with respect to the situation experienced. Resilience reflects a test that the entity or object must endure under a certain pressure, which implies succumbing or surviving. Hence, in the latter case, its quality of survival is binding with that of "hardness" or greater resilience, which ultimately, once the test phase is passed, can be translated into an entity or being those resists. By comparison, resistance is the capacity to face the pressure, confronting the force and avoiding impact by it. In a way, any object or living entity under any pressure by returning to its initial stage will demonstrate its resilience capacity which implies a form of resisting not receding entirely to that pressure. Consequently, resilience and resistance, although not synonymous, are two phenomena that correspond to the same path or trajectory.

A third term is heritage. It has a greater ancestry over time and according to records in the thirteenth century, the clergy was associated with inheritable assets. In Latin, heritage comes from *patrĭmōnĭum, ĭl.* (from *pater*). n. Cic. Heritage, family assets. // (fig.) Cic. Heritage, inheritance." (Fraile, 1984:346). The Oxford English Dictionary ratifies its link to "the history, traditions, buildings and objects that a country or society has had for many years and that are considered an important part of its character". (https://www.oxfordlearnersdictionaries.com/us/definition/english/resilience?q=resilience, reviewed eighth March 2020).

In recent decades, UNESCO has further qualified the term and its semantic coverage has transcended other areas, not without controversy and contradictions, to relate it mainly to culture and identity. One of the notions introduced is that heritage is not only a good or product but a process that is embedded in societies through their past, present and future.

The notion of heritage is important for culture and development as it constitutes the 'cultural capital' of contemporary societies. It contributes to the continuous revaluation of cultures and identities and is an important vehicle for the transmission of experiences, skills and knowledge between generations. It is also a source of inspiration for creativity and innovation, generated by contemporary and future cultural products. Cultural heritage holds the potential to promote access to and enjoyment of cultural diversity. It can also enrich social capital by creating a sense of belonging, individual and collective, that helps maintain social and territorial cohesion. On the other hand, cultural heritage has acquired great economic importance for the tourism sector in many countries, at the same time that new challenges were generated for its conservation. (http://www.unesco.org/new/es/mexico/work-areas/culture/intangible-heritage/, reviewed 13th December 2019, own translation).

In the first conceptual range of assets, based on the implementation of the Convention for the Safeguarding of Heritage (2003), the UNESCO established that heritage can be understood as material, natural and immaterial. In the latter case, it is pointed out that all heritage "must be safeguarded and consists in the recognition of the uses, representations, expressions, knowledge and techniques transmitted from generation to generation and that instill in communities and groups a feeling of identity and continuity, thus contributing to promoting respect for cultural diversity and human creativity" (http://www.unesco.org/new/es/mexico/work-areas/culture/intangible-heritage/, reviewed 13th December 2019, own translation). In particular, the manifestations of intangible cultural heritage can be appreciated in domains such as: traditions and oral expressions, including language as a vehicle for intangible cultural heritage; the performing arts; the social uses, rituals and festive acts; knowledge and uses related to nature and the universe and traditional ancestral techniques.

For the purposes of this chapter, we wish to draw attention to this type of heritage which specifically links "oral traditions and expressions (…) social uses (…) knowledge and uses related to nature and the universe". The narratives and experiences presented in the case of the *Eperara Siapidara* are an example of cultural heritage that reveals how events where resilience and resistance meet with certain clarity are culturally elaborated and reworked.

In what follows, we draw upon narratives and events experienced by the authors, as well as an intercultural dialogue between two *Eperara Siapidara* leaders and a *mestizo* anthropologist, to approach the concept of resilience, resistance and heritage — not so much in their meanings and trajectories examined here but rather throughout the *Eperara Siapidara* world. It is an intercultural exercise that seeks to examine and expand the conceptions and interpretations associated with the concepts of this triad. The contribution of this experience is to decentralize the notion of resilience that falls on the subject experiencing the disturbing pressure or action who ends up taking action of "withdrawing" or "turning back", as it is indicated by Latin meaning. It is clear in the examples that we will illustrate that, through narratives and experiences, the concept of resilience is coupled with that of resistance. We begin in a first section with a context section on the *Eperara Siapidara*, and then, in a second segment we delve into the narratives and events that associate resilience-resistance and heritage, and in the third and final section we present some conclusions about this interrelation which will surely have parallels in other cultural latitudes and certainly it would have some resonance to other contributions in this volume.

5.2 The *Eperara Siapidara:* Introducing the Context

According to the 2018 National Census, the *Eperara Siapidara* constitute one of the 115 ethnic groups that exist in Colombia. They are mainly located on the Pacific Coast of the departments of Cauca, Nariño and Valle. In Cauca they are established on the banks of the Saija River, municipality of López de Micay, and on the Naya River, municipalities of Tambo and Buenaventura, departments of Valle and Cauca; populations are also reported in the municipalities of El Charco and Olaya Herrera in the department of Nariño. The population for 2005 was estimated at approximately 3853; for 2018 the number rises to 7047 distributed in the Cauca, Nariño and Valle departments. (file:///C:/Archivos%20JET/Unicauca%202,020/presentacion-grupos-etnicos-2019.pdf, reviewed on 17th February 2020).

Although these departments are geographical references of the State Administration, for the *Eperara Siapidara*, these territorial borders do not exist, and although their displacement has occurred in the Pacific tropical forest ecosystem (Bernal, 2004), in recent years many inhabitants have moved to urban centres, young people among them, in search of opportunities that bring them closer to the dominant national system. In Jacinto's terms: "The indigenous person is where there is food, and there is living. We arrived in Guapi because there was food".

The native language is *Sia epẽrã pedée*, or *pede* ("the language of the people") which is associated with the Chocó macro-linguistic family. It is estimated that in pre-colonization periods by the Spaniards, the Eperara were identified as Chocó indigenous people, and that they presently have common features in cultural terms with other groups in the Colombian Pacific such as the *Emberá*.

According to the *Instituto Agustín Coddazzi*—IGAC, the Eperara Siapidara includes the groups Saija, Epená Saija, Epera Pedeé and Cholo (Codazzi, 2008:137). In Romilio and Jacinto's interpretation, there are three more that survived the unfortunate times that lie ahead of the Eperara Siapidara: those located in the *Nazarenes* of the Peña del Río Saija, San Francisco de Guanguí and those of the Río Micay. In their cultural tradition, there are several superior beings who accompany them in their itinerant life and who help them to resist in dialogue with nature: *Tachinawe* (Mother associated with the Moon), *Tachi ăk'ore* (Father associated with the Sun) and *Tachi eufa* (Mother Earth). These gods and goddesses are invoked before commencing agricultural, fishing and hunting activities that take place at various points along the coast in the departments of Cauca and Nariño, known as Gurupi, Guare, La Bocana, and Playa Blanca. There are other extractive activities, questioned by many, such as the indiscriminate cutting of wood—which incidentally constitutes one of the best-paid tasks that call into question their worldview of respect and protection for nature's resources. Women are also involved in artisanal activities based on basketry (Unidas, 2016). In summary, the Eperara Siapidara have a traditional knowledge of the ecosystem, enabling them to adapt to the various conditions they are faced with, which in turn also determines their resilience in the face of unexpected situations, both in urban and rural settings.

In the examples we wish to highlight, it is recognized that the elders have lived a long time and some oral or narrative traditions have been maintained over the years. Some cosmological versions of these traditions institutionalized through the National Indigenous Organization (ONIC) indicate that the *Eperara* conceive of two vital spaces. However, the version put forward by Jacinto and Romilio suggests three spaces are recognized: above (*Tachi ăk'ore*) where the spirits and gods reside, the current space where the riverine communities (*patri eufa p'anapata*) now live, and the world below, inhabited by others (*anta ăra'moră—tapanos*).[1] Along with these conceptions, there are others that, for our purposes, demonstrate the value of the knowledge and the *Eperara*'s narrative as an intangible heritage that resonates with the concepts of resistance and resilience. The *Eperara Siapidara*, like many indigenous peoples, possess an extensive richness of oral tradition about their ecosystem, their environment, and their socio-cultural interpretation of events that are presented to them in their daily lives and history. Here, we present only an illustrated sample of this wealth of ecosystem knowledge, with a specific connection associated with resilience and resistance.

[1] The three worlds are also related to various indigenous peoples not only in Colombia but on a planetary level. The interrelationships between the three are conceived as a whole that, as in Complex Systems Theory, maintain fluid and synergistic relationships within and outside the system.

5.2.1 *Shaaree na Reujade Panapata [Living the Force of our Territory with the Mighty Power—Vivimos la Fuerza de Nuestro Territorio Con el Poder Superior]: Approaches to Resilience and Resistance*

As previously indicated, intangible heritage involves oral expressions and knowledge of the environment that derive from adaptations of the past and that transcend through generations. In what follows, we will address two narratives of the past associated with the period of Spanish colonization experienced by the *Eperara Siapidara* and the description of an event lived by Romilio regarding the presence of paramilitaries in his territory. Both examples—of which there are many more, not necessarily associated with this specific narrative—reveal the concepts of resilience and resistance in a dynamic that articulates their knowledge of their environment that is assessed when foreign actors have a presence in their territory. A comparative analysis of the two stories is presented at the end of the text.

5.2.2 *Confronting the* Eperara Siapidara *Massacre: Resistance and Resilience in the Myth of Spanish Colonization*[2]

"According to the history, as indigenous peoples" says Jacinto, "we are the first. In the greatness of beings, the Eperara are the first Aboriginal people to dominate nature. That is the strength. We are the ones who predominate in these lands. *Tachi ăk'ore* warns us and tells us that this is going to happen, an earthquake, a flood and tells us 'buddy, see go meet at the Nazareno de la Peña' which is what will be left, everything else will end badly by the ill influence". This catastrophic antecedent led the conversation to one of the most mournful events in the history of the *Eperara Siapidara* and of America: the process of Spanish colonization.

According to Jacinto, the story was told to him by a grandmother,[3] who said that the indigenous people were subdued at a point known as 'the mouths of Patía' where they killed many indigenous people. The signs are still visible today: ditches where the blood of the murdered indigenous people passed through remain intact. It is said that the ditch was made by "the force of the people's blood". The story is told by

[2] In the past decades, the expression 'myth' has undergone transformations and deformations in its use. It is common to hear in various media "that's a myth" to indicate that it is a fictitious or falsified event. Despite the fact that anthropology as a discipline has demonstrated the interpretative value of what is understood by myth decades ago, this type of distortion still prevails. For the purposes of this text, we have chosen to assume a basic definition proposed by the anthropologist E. R. Leach: "a myth is a particular tale about the past which serves to justify a particular type of action in the present within the context of a particular cultural milieu" (Leach, 2000:106).

[3] According to Romilio, the story was told to him by his mother and her father told it to her, which indicates that the narrative has been transmitted for several generations.

a woman who, faced with this oppressive reality, said to her husband, "Why not do something? We are letting ourselves be killed. Could it be that we haven't learned anything about the art of witches, from the *jaipanás* of before, who were good? The idea was that the witch could kill the *cabezón* ("big head"), the leader of the Spanish group at lunchtime. They sent the leader a spirit and he vomited, and then said "No, let the indigenous people know that something is happening." Romilio complements the story by adding that in the plot, the witch sent a spirit in the water to the *cabezón* and when he drunk it, he got stuck. At that moment, the witch said, "I am going to send him a lightning bolt" and he got mad. To that, the Spanish leader said "No, let's go [release] these indigenous people because they 'are children of God' too, for there will be a reason this is happening to me". In this example, Romilio says, there is resilience because "there is still someone to protect you, a force." But before that event happened, he points out, there were already signs in nature—birds and animals warned. For example, little birds, tied each other with their wings, came out before the indigenous people as a sign of prevention. The elders were wise observers of nature and said that this was what was going to happen to them. Under pressure from the Spaniards, the cultural knowledge of the environment held by the *Siapidara* and their reputation as good witches reveal themselves as channels through which to return to their own cause and way of living.

5.2.3 "I live here": Resistance and Resilience to the Paramilitary Pressure

Paramilitary groups emerged in Colombia at the end of the 1980s when guerrillas were taking control in certain cattle ranch areas and drug dealers were developing their business in some regions. The paramilitaries then emerged by reaction to the presence of guerrillas in many territories of the country (Safford, 2002). In recent years, the Colombian pacific coast has become one of these macro regions affected by drug dealing, gold mining, guerrillas and paramilitary groups.

In 2015, in the Guare stream (a small stream of slow-moving water that flows into the Guapi river), where a group of *Eperara Siapidara* lives—an area near the head of the Guapi township—the paramilitaries entered the territory. Romilio lives there and his survival depends on fishing and hunting 'rabbits', *guagua* or *boruga* (*Cuniculus paca*). On one occasion, Romilio, or "Tovar" as his friends call him, and an Afro neighbour, tell him to go "*lamparear*"—hunting at night with a flashlight— and on this occasion, he was going out when friends told him that there was a group of people "quite heavily armed" and they were scared. "Tovar" asked them, "Why are you afraid if you have not done anything wrong?" to which they replied: "Tovar, since you are the one who knows, we put the faith in you that if we suddenly meet them you talk to them there". So, he conceded and at night he simply boarded his *canalete* (a small rowing boat) without knowing the exact point at which the paramilitaries were located. On the way down the stream, he suddenly detected a *ramada,* or shack, on the shore of a small beach. At this point, the paramilitaries called: "Hey, who is this?"

- "I" Tovar replied
- "Where are you going?" a paramilitary said, to which Romilio answered: "I'm going hunting."
- When Tovar said 'hunting', he heard that they de-secure their weapons.

Tovar continued further up the stream when he saw a 'rabbit' and *boom!* he hits it and as he was coming back down the stream, he was thinking, "Who could these guys be?" Then he thought, "Lord, let me pass over there, I know that with your power nothing will happen to me" [When questioning Romilio, Jairo asks him which Lord he thought of, to which Romilio replies, the "power of *Tachi ăk'ore*"].

Tovar told himself that with his power nothing will happen to him. Certainly, he passed by and they did nothing to him, nobody said anything. The next day, he said to himself, "I'm going to go see those people, whoever they were. I came back, I came back by day with a brother-in-law to see who those people were. He carried a box of bullets with the shotgun in the canoe. I said to my brother-in-law, 'They must be up there, those people.' And of course, when we got closer, my brother-in-law got scared and said: 'Oh my God, those armed people, we better get back!' I said, "No. If we lie down, they give us lead [shoot] because they will say they have come to (…)". And as we passed, they asked us:

- Where are you going?
- We said, 'fishing'.
- Buddy, do me a favour and come closer' they asked us.
- I arrived and they asked me 'where do you live?'.
- I live here in this Guare river, here below is my house.
- What do you live on?
- I told him about hunting, fishing and planting.
- How many years have you been here?
- I don't know, I replied.
- What about that little shot you have there?
- Well, they are mine because whenever I go hunting, I shoot the animal or people who come across me, and I go ahead giving them.
- When I told them this, they started laughing.
- So, they said, "just pretend you haven't seen anything".
- I replied, 'Of course, I don't know who you are, I have nothing to do with the life of *nadie* [nobody], only my life, because I know that my life whether I die or they kill me is worth a lot'.

Then we continued up the stream until we saw a turkey hen, and as I took out the shotgun, my brother-in-law said 'No, don't shoot, those people are down there.'

- To which I replied, "I don't have anything to do with them". So, I killed the turkey hen and we went back down, we brought fish too. If they call me, we give them fish. But when you go to talk to these people you have to go prepared with God.

- I had a box of cigarettes, they asked me for a cigarette and I gave them one, and since they didn't have any, they took it from each other. I had a leaf from a plant that I conjure and I carry it in a cigarette and the effect is that nobody can detains me.
- When I got home, I warned the family: "There are people who are armed, so you have to be prepared." It was 8 o'clock at night when an engine passed by.
- I said: "That engine that was a 40 (horse power) goes where those people are and with that they will return. When later the engine came back and I told my family to get ready, to make sure they were well placed on both sides of the stream.
- When the boat arrived, from there they called me; "Tovar!" and I heard the voice of an acquaintance and he told me that he wanted a favour to give him a gallon of gasoline because they were carrying a patient with a cape and a gun. The biker later told us that the paramilitary told him that if the indigenous people did not give him the fuel, they would kill them and take the fuel away. Then I told to our neighbours' Afros that after fifteen days these people should leave, and after fifteen days, they just left. I asked the one from above and he who is our police acted for us. The Afros said to me, "Tovar, you certainly know."

5.3 Intertextual Translation

What do these two narratives teach us? In dialogue with Romilio and Jacinto, we carried out an analysis attempting to connect the idea of intangible heritage, understood in this case as knowledge of the environment found in narratives as a variant of their ways of knowing, and the concepts of resilience-resistance.

First of all, in the first narrative associated with a history of the colonization period, the typical characteristics of a myth that structurally present opposite pairs, "they" (the Spanish) and "we" (*Eperara Siapidara*) different versions appear complementing each other. Jacinto's narrative was expanded by Romilio. In the encounter with the colonizers, the oppressive force assessed resistance and resilience, understood as the capacity of the *Eperara* to survive an oppressive event. The knowledge of the environment, embodied in their spiritual leaders like the *jaipaná*, enabled them to stop the violence of the invaders through the work of the spirits commanded by the leaders themselves. But the value of this action reveals another important interpretation that relates to the similarity or coexistence with the other as a form of adaptation. When the Spanish leader (*cabezón* or 'big head') gets stuck and is killed with a lightning bolt that knocks him out of his mind or out of his position, it puts him in a similar relationship with the other, the indigenous: ("No, let go of [those] indigenous because they 'are children of God' too, for some reason this is happening to me"). In a way, the attack action of the *jaipaná* generated a change of attitude in the other, in this case the Spanish leader, making him back down—resilience as "bouncing back"—and this allowed him to understand that the massacre of indigenous people should stop. In a way, the resistance of the *Eperara* became a reaction of the Spanish leader, "jumping or retreating" (due to the blockage), by accepting that the indigenous people are also children of God and therefore their livelihoods should be respected.

Second, the history associated with the event with the paramilitaries also reveals a dialogical relationship between resilience and resistance. In the first instance, the value of facing or confronting a threat and oppressive force telling the truth, understood in this case as the explanation, the reason for being or being in the territory, constitutes a key foundation of resistance. Romilio points out that it is possible for the paramilitaries to investigate who the real locals are and with that, when telling the "truth" there is no reason to fear because they can verify the information. With this decision, there is no fear, no lie—typical qualities of the armed conflict that is lived in conflictual territories in Colombia. One always has the option to lie, but in terms of resilience, this would not be a good choice. The other element of the world above is that Romilio entrusted himself to a superior God or *Tachi ăk'ore* and, in his spirituality, he had the conviction to ask for the call so that the meeting would be allowed and that later they would also leave for a new call. As in the previous myth, the connection with a superior force is verified, but this time by assertively telling the truth about belonging to a territory, the paramilitaries eventually understand the situation and let them continue their journey. Further, the knowledge of plants allows Romilio to dissuade the opponent from a requisition or mistrust of his presence in the area.[4] As in the previous narrative, resistance in the *Eperara*, facing them, standing firmly, dissuade the paramilitaries, backing down and leaving the territory as it was in its initial state.

These two narratives, distanced in time, teach us how the *Eperara Siapidara* confront and face foreign forces that threaten their stability and permanence in the territory. There are other events and conditions which, like these two narratives, reveal the connection between knowledge, resilience and resistance, but for now, this is just a sample of how, culturally, this Pacific people creatively resolve to threaten and destabilizing situations.

5.4 Some Learning Lessons and Conclusions

Resilience, resistance and intangible heritage are interconnected ideas. When preliminary discussions about the exercise began, the notions and genealogies of these ideas were at the forefront. Our argument in this chapter is presented as an intercultural exercise departing from a brief framework in the life history of the *Eperara Siapidara,* one of the 115 acknowledged indigenous groups in Colombia. The context of the Colombian pacific coast, in which the *Eperara Siapidara* live, allows us to

[4] In many indigenous societies the traditional healer uses to work against foreign forces such as illegal or legal armed forces who invade their territories by doing spiritual work. This spiritual resistance places another dimension difficult to describe as they form part of the strategies of resisting those who may cause damage in their territories.

examine the connection of these three ideas through their lived experiences in history, narrative and local knowledge.

In this intercultural exercise, we have learned several lessons. First, by approaching the historical narrative of the colonial encounter we found the notion of resilience in the strategies used to confront the invaders forces.

Thus, the knowledge of plants and spiritual healing appears as intangible heritage that is revealed in the conflictual tensions between the actors. Here, we have focused on narratives and tales, but it is certainly possible to conceive of other expressions, events or manifestations that are also connected to this concept. This provides motivation to consider other dimensions of social life where these concepts can be examined.

Second, resilience does not precede resistance, and conversely, resistance does not precede resilience. Both are coextensive and it is prudent to regard these concepts as a *continuum*, although some authors and institutions perceive them as a process rather than a product (International 2016). This leaves the field open to inquiry, especially in the interrelation with other oppressive, stressful, or threatening events for the populations under enquiry. In this regard, memory and history play a crucial role in perceiving the common thread of this *continuum* as revealed by the myth of the colonial period in the *Eperara*.

Third, the collaborative and self-reflection methodology amongst two *Eperara* and one *mestizo* has proven to be a productive exercise in examining cases of resilience and resistance in social groups or indigenous communities by using a method of enquiry proper of the social disciplines. The exercise requires further polishing, but it shows promising results, as previous works on collaborative research have demonstrated (Borda, 1986; Selener, 1997; Borda, 2001; Rappaport, 2005a; Rappaport, 2005b; Vasco, 2007; Tocancipa-Falla, 2010). Further, it must be said that, to a large extent, this perspective is a masculine version of how these concepts interrelate; the feminine gender perspective of the events would have to be examined, although the first narrative was evoked by a grandmother.

Finally, the triad of knowledge as immaterial heritage, resilience and resistance, and vice versa, opens various windows to the social world where its interrelationships and connections can be identified. There are pending issues to be addressed in future research; for now, this is a first approach, and simultaneously an invitation to maintain intercultural discussion and dialogue in order to continue reaping lessons that enhance understanding and extension to other latitudes.

References

Bernal, E. H. (2004). Naturaleza y territorio. Cosmovisi6n, sistemas productivos y medicina tradicional Eperara Siapidara. Bogotá, Gente Nueva Editorial.

Borda, O. F. (1986). Conocimiento y poder popular. Bogotá, Punta de Lanza/Siglo Veintiuno Editores.

Borda, O. F. (2001). Participatory (action) research in social theory: origins and challenges.In P. Reason, & H. Bradbury (Eds.), *Handbook of action research: participative inquiry and practice*: 27–37.
Codazzi, I. G. A. (2008). Atlas Básico de Colombia, Tomos I y II. Bogotá, Imprenta Nacional.
Española, R. A. (2001). Diccionario de la Lengua Española. España, Real Academia Española.
Española (2018). https://dle.rae.es/resiliencia
Fraile, A. B. (1984). Diccionario Manual Latino-Español y Español-Latino. Barcelona, Ramon Sopena S.A.
Geertz, C. (1983, 2000). Local knowledge. New York, Basic Books.
Kotliarenco, M. A., et al. (1997). Estado de Arte en Resiliencia. s.f., Organización Panamericana de la Salud/Oficina Sanitaria Panamericana, Oficina Regional de la Organización Mundial de la Salud/Fundación W.K.Kellogg/Agencia Sueca de Cooperación Internacional para el Desarrollo.
Leach, E. R. (2000). 1.9 Claude Levi-Strauss: Anthropologist and Philosopher (1965). In S. Hugh-Jones & J. Laidlaw (Eds.), *The essential edmund leach. Volume I: Anthropology and society* (pp. 97–112). Yale University Press.
Mora, J. F. (1994). Diccionario de Filosofía. Barcelona, Ariel.
Rappaport, J. (2005a). *Intercultural utopias: public intellectuals, cultural experimentation and ethnic pluralism in Colombia.* Duke University Press.
Rappaport, J. (2005b). Retornando la mirada: una investigación colaborativa interétnica sobre el Cauca a la entrada del milenio. Popayán, Universidad del Cauca.
Safford, F., Palacios, M. (2002). Colombia: Fragmented land, divided society..
Selener, D. (1997). *Participatory Action Research and social change.* Cornell Participatory Action Research Network.
Tocancipa-Falla, J. (2010). El mundo de la vida, crisis y etnografía: aproximaciones antropológicas a la fenomenología de Husserl. Anuario Colombiano de Fenomenología IV, 199–217.
Unidas, E. (2016). Cultura Eperara Siapidaara. Bogotá, Fundación Sura/Fundación Etnollano.
Vasco, L. G. (2007). Así es mi método en etnografía. Tabula Rasa, *6*, 19–52.
Werner, E. E. (1989). High-risk children in young adulthood: a longitudinal study from birth to 32 years. *American Journal of Orthopsychiatry, 59*(1), 72–81.

Chapter 6
Adaptative Model Based on the Zenú Culture

Rafael Espinosa Forero, Lina García-Corrales, Carlos Castaño Uribe, and Wilfredo González Martínez

Abstract The Zenú adaptation system was one of the largest prehispanic landscape modifications in Colombia. It comprised a series of *camellones* and canals used for home, for agricultural and fish production in a resilient way against variations of water levels. The purpose of this project is to study the wetland use and exploitation, based on adaptation technologies used by the Zenú natives. The adaptation model is located in an urban area in the Purísima municipality next to the wetland. The methodology includes planning and implementation of an open and semi-open hydraulic system and agro-ecological, restoration and hydrobiological and faunistic management protocols. Results reveal successful traditional agricultural and fish species production, the wetland ecological restoration, resilience improvement against flooding and drought and the intangible Zenú cultural heritage strengthening. This evidence grants the replication of the model in water body neighbouring communities.

Keywords Adaptation · Zenú · Cultural heritage · Wetland

R. E. Forero
Regional Autonomous Corporation of the Sinú and San Jorge Valleys—CVS, Montería, Colombia
e-mail: rafael.espinosa@cvs.gov.co

L. García-Corrales (✉)
Department of Civil and Environmental Engineering, Universidad del Norte, Barranquilla, Colombia
e-mail: lmcorrales@uninorte.edu.co

C. C. Uribe · W. G. Martínez
Caribbean Environmental Heritage Foundation, Santa Marta, Colombia
e-mail: direccioncientifica@herenciaambiental.org

J. Morato et al. (eds.), *Intangible Cultural Heritage and Traditional Knowledge Related to Urban Resilience*, Resilient Cities, https://doi.org/10.1007/978-3-032-12713-6_6

6.1 Introduction: Cultural and Ecological Context

6.1.1 Zenú Culture Adaptation Practices

The ancient Zenú territory was home for this culture for about 2000 years, since 800 BC–1200 AD. The Zenú inhabited the Sinú, San Jorge, Cauca and Nechí river watersheds, which is an area with plenty of rivers, channels and swamps. The Zenú territory was divided into three provinces: Finzenú, Panzenú and Zenúfana. The Sinú river valley belongs to the Finzenú province (Sánchez Jabba, 2013). At Spanish arrival, the important Zenú populations, who lived in the Sinú Valley, were not located on riversides, as nowadays, but close to the swamps (Fundación Herencia Ambiental Caribe & CVS, 2014).

Studies made by several investigators, especially from the Museo del Oro, shows no evidence of prehispanic settlements in zones that are not subjected to flooding, and the settling pattern on swampy areas shows occupation of flood areas by means of land modification (Montejo Gaitán & Rojas-Mora, 1992). The community system was comprised of large networks with shape and function diverse canals and *camellones* that granted production and provided artificial platforms used as living places (De Fontainieu, 2006). *Camellones* were used to cultivate the soil with nutrients resulting from sediments and fishing activity canals, which took advantage of natural hydrological flows, sediments and native fish species. The management model supposed the definition of a production process granting the community's adaptation to the repeated drought and flooding conditions (Fundación Herencia Ambiental Caribe & CVS, 2013).

The pre-Columbian population in the Sinú valley exploited the regions' geographical characteristics by building sophisticated *camellón* systems in order to canalise, drain and connect water bodies as a response to the changing conditions in flooding levels (Banco de la República, 2010). During rainy seasons, the *camellones* contained floods and helped with the natural operation of swamps and protected crops and homes from overflowing waters. Its design diminished the speed of water and contained the river's sediments, thus, in dry seasons, they could be used as compost for the Zenú's farming areas (Fundación Herencia Ambiental Caribe & CVS, 2014).

It is estimated that the area covered by this system in the Sinú watershed is of about 150.000 ha and is considered as one of the largest landscape modifications made by prehispanic populations in Colombia (CVS & Universidad Nacional de Colombia, 2007). Artificial canals and *camellones* have been observed in aerial photographs on the Sinú middle and lower valley. Figure 6.1 shows a Zenú hydraulic system example.

Cultural background and ancestral elements are present in Zenú descendants, which represent an essential basis to reconstruct part of the adaptation model. Nowadays, elements such as the following are still present: fishing and wildlife exploitation techniques, agro-ecological arrangements, knowledge of hydrological flow management and its interrelationship with fish species cycles in the Sinú fluvial system (Fundación Herencia Ambiental Caribe & CVS, 2013).

Fig. 6.1 Current appearance of the ancient Zenú artificial canals and camellones. *Source* Banco de la República (2015)

6.1.2 Ecological System: Lower Sinú's Swampy System

The Lower Sinú's Swampy System (CCBS as the initials in Spanish) is a wetland system declared, in 2007, as sustainable use protected area. It is located in the northern part of the Córdoba department in Colombia, on the Sinú river's right edge, Fig. 6.2 (CVS & Universidad Nacional de Colombia, 2005).

The complex is an environmental system composed by an intricate network of channels, minor depressions and swamps. This system is fed by the overflow produced by its own basin and the Sinú river's flows. The complex's main hydrological function is to naturally balance and regulate both the local rain and the Sinú river's excess water volumes (CVS & Universidad Nacional de Colombia, 2007).

The hydrographic basin comprises an area of 3140 km^2 approximately. The maximum flooding area of the lower Sinú's swampy complex is of about 440 km^2 (CVS & Universidad Nacional de Colombia, 2005). Water levels present an annual uni-modal cycle and the variability of the swampy complex's flooding levels result from the local weather conditions and the Sinú river's hydrology. A low water level season between February and April and high-water level season between August and October are present. Transition from high to low water levels and vice versa takes place from November to January and May to July respectively. In average hydrological conditions, in low water level seasons, the flooded area is reduced to 15% of

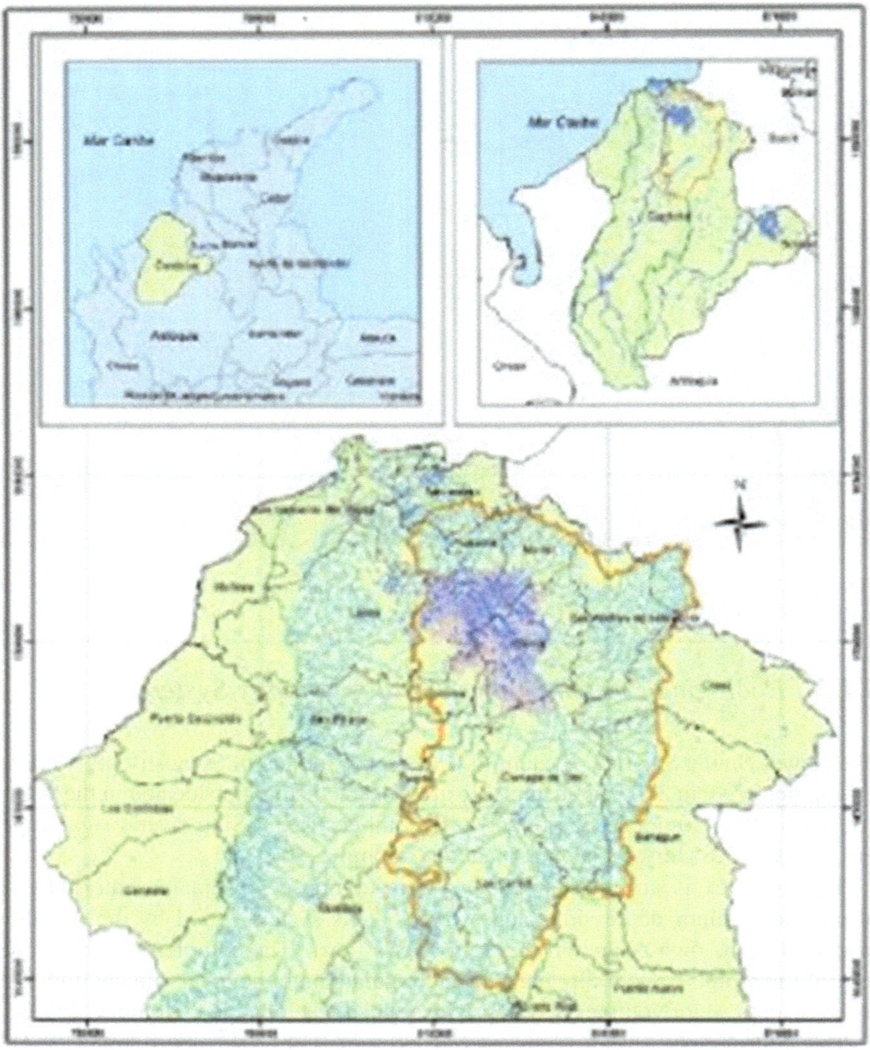

Fig. 6.2 The Lower Sinú Swampy System location. *Source* CVS & Universidad Nacional de Colombia (2005)

the whole area (CVS & Universidad Nacional de Colombia, 2007). Flooding movement and variability grant the development of several productive activities according to the season.

The swampy system depends not only on its exchanges with the natural environment where it is located, it also depends on the interrelations held with its social environment, the land's economic use in this area related to its possessing structure, the small landholdings dedicated to traditional or survival agriculture and large landholdings dedicated to extensive stockbreeding and commercial agriculture. Un-suitable production techniques create a series of impacts to the ecosystem and changes its hydrological and ecological dynamic (CVS & Universidad Nacional de Colombia, 2007).

Currently, the complex is influenced by a series of human interventions related to land adaptation processes through dike construction, drain canals and pumping systems, which is a widespread practice that allows settling in flooding area. Technological agriculture and cattle grazing activities established inside the complex and in the surroundings generate drainage of the complex and loss of its swampy system functions (CVS & Universidad Nacional de Colombia, 2007).

6.2 Planning an Adaptative Model

6.2.1 Land Selection

In 2013, among the Lower Sinú Swampy Complex neighbouring communities, parcels of land were identified, with a potential for the Zenú adaptation model's application, taking into consideration that communities still hold some traditional practices such as fishing techniques, use of fibber materials for handcrafting, agrosytems for subsistence and ancestral knowledge of hydrological flows. At the beginning, pieces of land were identified from the following social organizations: Fundación Ciénaga de Bañó, Agrosistema familiar El Pescador Inés Chiquita and Asociación Comunitaria Apropapur. During the identification process, infrastructure improvement, organizational strengthening and production systems and eco-tourism educational activities were also executed (Fundación Herencia Ambiental Caribe & CVS, 2013).

The land parcel selection criteria included: Location, Sociocultural and community, Natural resources use, Existing infrastructure, Soil characteristics, Limnology. In Table 6.1, the land's evaluated criteria and its description are shown.

After evaluating the chosen pieces of land, the Asociación Productora de Patilla de Purísima's area was selected. It is located on the northern area of the Lower Sinú's Swampy Complex on the city of Purísima's limits. Its geographical coordinates are 9° 13′ 46″ N and 75° 43′ 30″ W. The land holds an area of 6.6 ha. When selecting the area, the land counted on about 3 ha of agricultural and fish farming production with ponds attached to the wetland. Some highlights in this land are knowledge of fish farming and agriculture, a collective production diagram, work commitment, a significant number of families (25), women's active participation, government program experience and storage areas. It is especially emphasized that association members take part in cultural celebrations and keep farming following ancestral customs, though fishing arts are not preserved. Moreover, Zenú pottery has been found close to the chosen land. At that moment, the community had low incomes, had problems selling their products and public services were poor (Fig. 6.3).

Table 6.1 Land selection criteria. *Source* Herencia Ambiental Caribe & CVS (2013)

Criteria	Description
Location	Proximity to the wetland, land area, access roads (water round of the wetland called "payon" or "playones" which is the site that shows this system that is generally exposed when the level of the water pulses of the swamp decrease. Peasants and indigenous people used these beaches to grow fast-growing products (cassava, patilla, among others)
Sociocultural and community	Organizational structure (number of families involved, abilities, training needs, project execution experience, community integration and neighbour communications, among other strong and weak aspects) Housing conditions (access to public services, housing characteristics, family members' roles) Cultural aspects (Participation in traditional festivities, ancestral customs, use of the territory, archaeological sites)
Natural resources use	Agriculture: cultivated species variety and farming techniques used. Fish farming: Species and fishing arts used by the Zenú, production registration and hydraulic system (closed, open or semi-open)
Existing infrastructure	Fish farming ponds' conditions, storage areas, inner paths
Limnology	Biological communities (phytoplankton, zooplankton, aquatic macro invertebrates and microphytes). Water's physical and chemical variables (temperature, dissolved oxygen, pH, etc.)
Soil characteristics	Agriculture: nitrogen, phosphorus, ions, cation exchange capacity, pH, among others. Fish farming: total amount of organic material, sulphur levels, acidity and Ph

6.2.2 Hydraulic System Definition

The system's operation used by the ancient Zenú relied substantially on seasonal hydrological flows known as open system, a system which is no longer used. An open system design is composed of canals, *camellones* and terraces, which grant semi-confinement of water and sediments and avoids rapid loss, which also promotes exploitation over longer periods of time without substantially interfering with the natural dynamics (Fundación Herencia Ambiental Caribe & CVS, 2013). Open system installation allows water circulation in order to retain a larger amount of humidity over the longest possible time and promotes water exchange to the main body of water's flow pattern by means of complex channel and *camellón* systems. This kind of system eases permanent and very productive intensive agricultural practices based on adaptation of irrigation, draining and sediment. The open systems promote high efficiency in food production and more effective sociocultural arrangements.

The current state of hydrological management projects for fish farming use of the wetland neighbouring communities is named as a Closed System. It comprises an artificial water body, in which water only flows into during flooding and must be externally supplied for water replacement. An alternative design is the Semi-Open

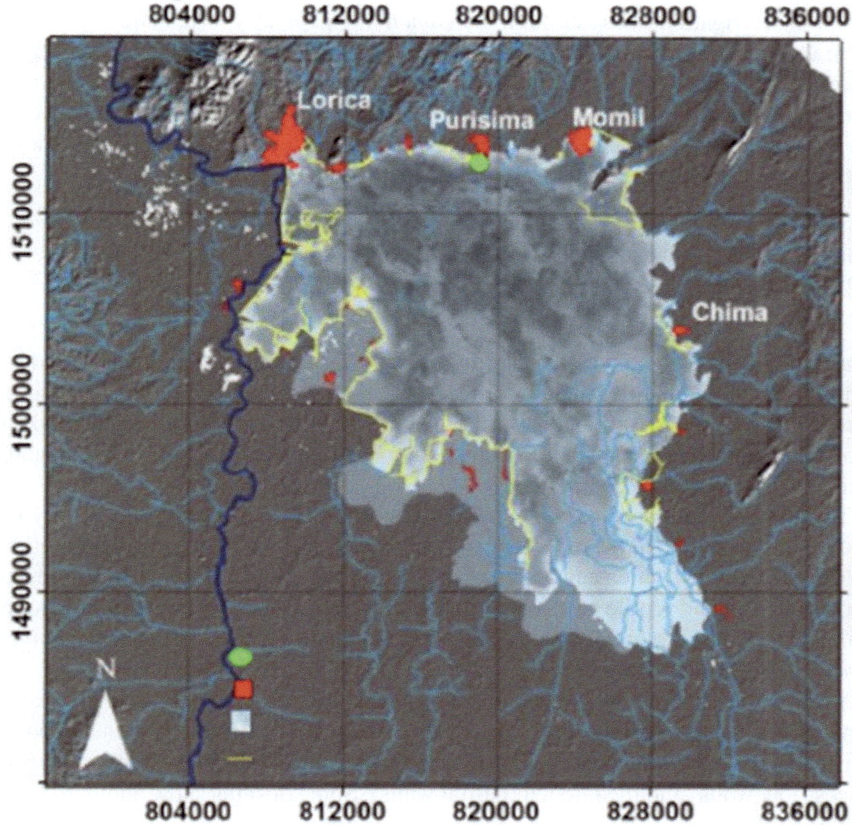

Fig. 6.3 Ubication of selected land. *Source* Fundación Herencia Ambiental Caribe & CVS (2013)

System. It uses floodgates in order to regulate water flows depending on flow conditions and it can be supplied by means of direct connection to the main hydrological system.

An open and semi-open system was designed, in order for the Zenú technology to be applied and the infrastructure of the property was used. The design parameters were topography, annual water level variability and precipitation. The analysis of water levels in the area was performed for the wetland sector, named as El Playon swamp (CVS & Universidad Nacional de Colombia, 2005). When creating the *camellón*'s elevation design, the maximum altitude was taken as a guide in order to avoid overflows by flooding events. In the same way, the minimum altitude was designated to be the level in which it is expected from the system to be capable of working without need of external water supply for irrigation.

In the model's design, the water level's seasonal cycle is also considered. During low water level months (December–March) the use of soil water is maximized in order to diminish irrigation needs through the *camellón*'s geometry, orientation and agricultural species disposition manipulation. During high water level months (April–March) excess is regulated mainly by selecting the agricultural species. The average annual precipitation in the complex is from 1200 to 1300 mm. The seasonal trend is very similar to the water level, the coincidence of low water levels and minor precipitation periods are some of the challenges of the design.

6.2.3 Characterization of Zenú Elements

Characterization of elements ancestrally used by the Zenú has been performed in homes in the municipalities of San Andrés de Sotavento, Tuchín, Chimá and Purísima (Fundación Herencia Ambiental Caribe & CVS, 2014). The characteristics taken into consideration in the adaptation model are: vegetation use, wildlife use, housing designs and cooking and work tools.

Twenty-two plant species rooted in the Zenú homes were registered, mostly of medicinal properties and a few domestic use species. Some of the more important are Corn (*zea mays*) which can be found in 13 varieties. Wildlife used corresponds to taxonomic bird and reptile groups that match the studies made by (Carvajal-Cogollo et al., 2007). They are mainly used as house pets; only morrocoy turtle (*Chelonoidis carbonaria*) is used as food.

Variation in architectonical elements is not very present and traditional constructions are made of materials such as canes in the walls and palm leaves as roofs. Home designs hold three spaces: dormitories (with walls rising to the ceiling), social spaces (low walls) and the kitchen (partial walls). Some designs show another space as storage (Banco de la República, 2010). Houses are surrounded by a spacious yard shared with family homes. Figure 6.4 shows an example of a Zenú home.

Among the kitchen tools, totumo tree fruit (*Crescentia cujete)*, use is extensive, which is turned into spoons, coffee cups, containers, strainers and handmade weighing scales. Fibbers derived from palm species such as the *Gynerium sagitatum)* or iraca *(Carludovica palmata)* are used for crafting of elements such as baskets and the *vueltiao* hat. Many other tools are made of clay, such as vessels and pots, which are not widely used. Other wooden elements are also frequent. Nowadays, some Zenú use modern plastic products as substitution for traditional elements.

6.2.4 Floristic, Faunistic and Ichthyology Characterization

The planning of the project also took into account the floristic, faunistic and ichthyology characterization in order to define the productive component one (Fundación Herencia Ambiental Caribe & CVS, 2014).

Floristic characterization included identification of Zenú ancestral plant species that are still used and those which are not. Plants that are still part of the Zenú culture

Fig. 6.4 Zenú home in Purísima. *Source* Fundación Herencia Ambiental Caribe & CVS (2014)

are those with agricultural production characteristics for family food. Some of them are yucca *(Manihot esculenta),* plantain (*Musa paradisiaca)*, corn and yam (*Musa paradisiaca)*, and also vegetables such as pepper (*Capsicum annuum*), tomatoes (*Capsicum annuum*), beans (*Phaseolus vulgaris*), runner beans (*Vigna unguiculata*) eggplant (*Solanum melongena*). Some of the species no longer cultivated are medicinal plants: arnica (*Arnica spp*), tres bolas (*Hiptis capitata*) and caraguala (*Disocactus amazonicus*).

In addition, floristic characterization of the chosen land was made. The most representative species are yucca, corn, runner beans, beans, pepper and yam. Avocado (*Persea americana*), mandarin (*Citrus reticulata*), bitter orange (*Citrus x aurantium*), elderberry (*Sambucus spp*), Spanish lime (*Melicoccus bijugatus*), cantaloupe (*Cucumis melo*), papaya *(Carica papaya),* watermelon (*Citrullus lanatus*) and cashew (*Anacardium occidentale*) were also detected.

Faunistic characterization was made on the project's land in taxonomic groups of wildlife, amphibians, reptiles, birds and mammals. Nine species of amphibians were identified, the Hylidae family being the most common. Eight species of reptiles were identified and the most common is the Teiidae family. 52 bird species, especially from the Thraupidae and Tyrannidae families were identified. Identified mammals include 5 species, being the Muridae the most predominant. Finally, native fish species identification included four species: yalúa (*Cyphocharax magdalenae*), cocobolo (*Aequidens pulcher*), tilapia (*Oreochromis niloticus*), yellow surfperch (*Caquetaia kraussii*), panchita (*Astianax fasciatus*) y eel (*Synbranchus marmoratus*).

6.2.5 Protocol Elaboration

Protocols were elaborated after choosing the land, defining the hydraulic system and characterizing the Zenú elements, flora, fauna and fish species (Fundación Herencia Ambiental Caribe & CVS, 2014). Three protocols were defined: agro-ecological system, ecological restoration and hydrobiological and faunistic element management. Currently, the model is in execution of the planned protocols.

The execution protocol of an agro-ecological system is aimed to apply agricultural techniques used by the Zenú, promote seed preservation, measure the species' productivity and increase the association's agricultural production and strengthen the community work in Apropapur. This protocol includes sowing designs on the *camellones*, sowing density and procedure, monitoring parameters after sowing and the farming conditions for agricultural products.

The ecological restoration protocol was designed as a strategy for preservation of the wetland's riparian vegetation. It is aimed to recover, create and offer habitat availability for wildlife, recover biotic interactions, improve ecological connectivity of floral communities and improve the landscape aesthetics. This project comprises an initial environmental analysis of an area to be protected, on weather stressors and native vegetation potential. Also, zoning management (restoration, preservation and use) was performed, and designs were elaborated containing selections of species and the floristic arrangement definition.

Finally, the faunistic and hydrobiological element management protocol aims at the evaluation of environmental advantages for *camellón* systems. The designed activities in the protocol are: community training, species identification and recognition of their characteristics (behaviour and implications for the project), habitat creation and availability by settling natural species (percha, forrajeo, shelter and feeding), information recording, weather stress management, and monitoring and evaluation of earlier activities.

6.3 Implementation of Adaptative Pilot Model

6.3.1 Construction of the Hydraulic System

As part of the Adaptation Model's implementation, 11 *camellones,* open system and semi-open water reservoirs were built, connecting two closed system reservoirs into the system and the definition of a protected area. In Fig. 6.3, a picture of the system's components is shown, black numbers identify *camellones.*

During 2004, construction of *camellones* interspersed by hydrological flow canals was performed; these camellones and canals were designed according to the area's characteristics. The *camellón*'s dimensions are shown in Table 6.1. Concurrently, reservoir 1 was built to maintain large water volumes and supply for agro-ecological and fishing requirements; allowing monitoring and experimentation with the incoming native fish species by natural hydrological flows (Fundación Herencia Ambiental Caribe & CVS, 2015) (Table 6.2).

In 2015, reservoir 2 was adapted and 1 m of depth in the area was added. Reservoirs 1 and 2 were also connected by installing a floodgate between them. Reservoirs 3 and 4 were adapted as connected semi-open systems and a floodgate was installed between reservoirs 2 and 3. The system's total storage volume is 14,112 m^3. It must be considered that 12,310 m^3 are comprised in the four reservoirs and the rest is located in the canals (Fundación Herencia Ambiental Caribe & CVS, 2015), Table 6.3 shows the reservoirs' dimensions.

Also, in 2015, restoration of the main canal of 280 m was done, which allows for water supply for the canals from the swamp to the model. In this way, the system remains interconnected both from the inside and to the swamp. In 2016, the *camellón* 11 was modified, resulting in the enlargement of the agricultural production system. Currently, the *camellón* is 145 m long and 15 m wide (Fundación Herencia Ambiental

Table 6.2 Camellones dimensions. *Source* Herencia Ambiental Caribe & CVS (2014)

Camellón	Length (m)	Width (m)	Height (m)
1	30.00	7.20	2.20
2	20.00	6.20	2.10
3	20.00	6.20	2.10
4	20.00	6.20	2.20
5	20.00	6.20	2.20
6	19.20	7.20	2.10
7	17.50	7.20	2.10
8	30.00	12.00	1.60–2.20 (variable)
9	20.00	12.00	2.10
10	43.00 × 107.00 (L form)	6.20	2.10
11	132.00	6.20	3.00
12	100.00	12.00	2.20
13	100.00	12.00	2.20
14	20.00	6.20	2.10
15	25.00	8.00	2.10
16	30.00	12.00	1.60–2.20 (variable)

Table 6.3 Reservoir dimensions. *Source* Herencia Ambiental Caribe & CVS (2015)

Reservoir	Area (m^2)	Volume (m^3)	Depth (m)
1	1687.8	2531.8	1.5
2	2846.7	5693.3	2.0
3	2304.4	3456.6	1.5
4	523.5	628.1	1.2

Caribe & CVS, 2016). Lake sediments were placed on top of the c *camellones* after water removal, which imitates the Zenú technique.

Since September 2016, rainfall and wetland levels have been monitored. The pluviometry devices were located at *camellones* 8 and 10, while the levels were taken with an artisan graduated staff at the entrance of the system next to reservoir 2. The results of these actions allowed us to relate the water conditions of the system with the protocol's implementation results.

6.3.2 Implementation of Protocols

6.3.2.1 Agro-ecological Protocol

The agro-ecological protocol allowed sowing of agricultural species used by the Zenú in the *camellones.* On the base and on top of the *camellones*, material from the Canutillo (*Hymenachne amplexicaulis*) and bocachica (*Thalia geniculata*) was placed as a protection strategy against erosion and humidity conservation. For each of the *camellones*, the design of the cultivating species, sowing density, sowing and harvest procedures was defined. On the *camellón*'s slopes, runner bean was planted and, on the top, plantain, yucca and corn were mainly planted. Also, fruit and native species in the parcel of land were planted (Fundación Herencia Ambiental Caribe & CVS, 2014). A list of fruit species can be found in Table 6.4.

Results derived from implementation of the agro-ecological protocol will be illustrated chronologically. Hence, in 2014, despite the rain decrease due to the El Niño phenomenon, successful harvesting of beans, watermelon, plantain, melon, yucca, cucumber and eggplant was achieved. In addition, relevant data for agroecological system management in dry conditions was gathered.

In 2015, several products were planted and harvested: runner bean, eggplant, melon, plantain, corn, yucca, sweet guava, sweet orange, lemon *(Citrus × aurantifolia)*, cucumber *(Cucumis sativus)*, passion fruit *(Passiflora edulis)*, papaya *(Carica papaya),* ahuyama (*Cucurbita maxima*) and pigeon pea (*Cajanus cajan).* Adaptation of yucca-yam, bean-yucca, pigeon pea-ahuyama, and passion fruit-papaya combined crops stands out with excellent results and a production of almost 1 ton of foodstuff. Higher levels of productivity were noted for yucca and pepper using sediments and

Table 6.4 Fruit species planted on the top of camellones. *Source* Herencia Ambiental Caribe & CVS (2014)

Common name	Scientific name	Common name	Scientific name
Mango	*Manguifera indica*	Sweet guava	*Psidium guajava*
Soursop	*Annona muricata*	Sour guava	*Psidium araca*
American-oil plant	*Elaeis oleífera*	Sweet orange	*Citrus x sinensis*
Watermelon	*Citrullus lanatus*	Cherry	*Prunus cerasus*

Fig. 6.5 Adaptative model components. *Source* Fundación Herencia Ambiental Caribe & CVS (2014)

Fig. 6.6 Corn, yucca and mango grown in camellón number 9. *Source* Fundación Herencia Ambiental Caribe & CVS (2015)

organic matter supports from the construction of canals and the wetland's bottom when water levels decreased (Fundación Herencia Ambiental Caribe & CVS, 2015) (Figs. 6.5 and 6.6).

In 2016, production in two *camellones* was compared: No. 8, enriched by sediments and organic matter, and No. 11, not enriched. Species selected for comparison were yucca, corn, pepper and bean. Germination times, leaf sprout, growing and development were monitored. Results show that enriched soils grant better vegetation conditions, in spite of the presence of salts in the *camellón*'s soil which revealed disparagement between the indicators. At this moment the production and harvesting monitoring continues.

6.3.2.2 Ecological Restoration Protocol

During the implementation of the ecological restoration protocol, an area of 248 m^2 was selected for riparian vegetation preservation, named No. 12 in Figure 6.5. Native species selected were: saman (*Samanea saman*) working as woody coverage

Fig. 6.7 Image of the project in 2017. *Source* Fundación Herencia Ambiental Caribe & CVS (2015)

whose roots assist erosion control, caracolí (*Anacardium excelsum*) a flooding and drought adapted species which is used as shelter for birds and mammals, and guadua (*Guadua angustifolia*) soil protection contributor and environment embellisher. Floristic designs are shown in Figure 6.7.

During the implementation of restoration protocol, monitoring of plant species was carried out. Excellent results were obtained for saman and caracolí, although no growth by the planted guadua species was observed due to weather condition in 2014. In compensation, guásimo (*Guazuma ulmifolia*), dorado (*Casearia sp*) and bajagua (*Senna reticulata*), which are native plants and were present in the area before the restoration activities, were preserved and added to the floral arrangement.

6.3.2.3 Faunistic and Hydrobiological Element Management Protocol

In relation to wildlife management, the protocol's implementation included information gathering of amphibians, reptiles, birds and mammals. A diversity of birds, newly come to the area, stands out, and iguanas are present as cultivation stressors.

Regarding the hydrobiological elements, several fishing days were carried out. Species yalua, wolf fish, tilapia, yellow surfperches, moncholo (*Hoplias malabaricus*), bocachico (*Prochilodus magdalenae*) y liseta (*Chelon labrosus*) (Fundación Herencia Ambiental Caribe & CVS, 2015), growth and natural feeding without human interference were identified. Natural feeding was achieved by allowing the entry of water and nutrients and confining fish when water levels drop. Moreover, the predominant presence of the bocachico species, which is part of the Zenú traditional eating habit, was highlighted. Figure 6.8 shows Ichthyological monitoring.

During low water level periods and lesser rain in 2014 and 2015 (accompanied by the El Niño phenomenon), reservoir 1 allowed fishing exploitation. Only reservoir 1 kept its water level and allowed for the endurance of several species that entered through hydrological flows. This reveals the system's effectiveness.

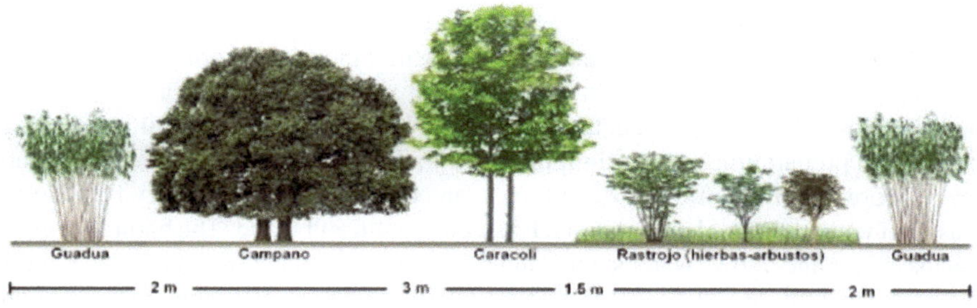

Fig. 6.8 Floristic arrangement design of protected area. *Source* Fundación Herencia Ambiental Caribe & CVS (2014)

The results from Ichthyological monitoring suggest that the proposed model facilitates a greater gain in size and weight of the fish compared to the conventional closed system. The latter reveals the capacity of the model for adaptation to changing climatic conditions.

6.3.3 Zenú Elements Appropriation

Zenú elements were incorporated into fishing, planting and harvesting activities, as part of adaptive model implementation, and have allowed a greater appropriation by the community. The Zenú elements appropriation and the training activities turned out to be an attractive aspect for cultural tourism and ecotourism development, and scientific and educational institutions' visits. For example, by FAO (Food and Agriculture Organization) and Colciencias (science, technology and innovation public institution of Colombia) (Fundación Herencia Ambiental Caribe & CVS, 2016).

In 2015, an area destined to be used for the museum establishment—Zenú culture interpretative room—was added and aimed to exhibit the adaptation model and show recovered archaeological elements. In addition, a kitchen and dining room were built based on the Zenú architecture, which also allows for agricultural and fishing production preparation and consumption with traditional utensils, especially totumo tools. Moreover, in relation to Zenú topics, an ecological bathroom (dry bathroom) was built with traditional materials such as canes, palm and wood. The bathroom does not use water and allows water saving and a source for compost (Fundación Herencia Ambiental Caribe & CVS, 2015). Figure 6.9 shows Zenú elements appropriation (Fig. 6.10).

Fig. 6.9 Ichthyological monitoring. b. Measurement of bocachico individual. *Source* Fundación Herencia Ambiental Caribe & CVS (2015)

Fig. 6.10 Zenú elements Appropriation. *Source* Fundación Herencia Ambiental Caribe & CVS (2016)

6.4 Conclusions

The adaptive model for the wetland's neighbouring communities was designed based on intangible cultural heritage and traditional knowledge of the Zenú culture, which still prevails in inhabitants in zones close to the Lower Sinú's Swampy Complex on the Zenú ancient territories. The Zenú adaptation community system included a vast network of canals and *camellones* that granted its productive exploitation. In this adaptation model, ancestral practices were applied, consisting of sowing on top of the *camellones* with soils enriched by nutrients resulting from the sediments. Additionally, natural hydrological flows were used and native fish species inflow was permitted for fish production as the Zenú did.

The adaptation model based on the Zenú culture was implemented in a land zone next to the wetland, located on the Purísima's urban area. The planning process included land selection stages, definition of a hydraulic system, floristic, faunistic, hydrobiological and cultural element identification, and creation of implementation protocols. This represents a course for future replication of the model.

The construction was made in stages and includes 11 *camellones* and 4 water reservoirs which allowed for implementation of the designed protocols. The results of the agro-ecological protocol reveal that, even in poor rain and water level conditions in the wetland, due to the influence of the El Niño phenomenon, it was possible to successfully harvest various traditional products. It demonstrates the model's potential for water use maximization and adaptation to changing ecological conditions.

The implementation of restoration protocol allowed for conservation of a riparian vegetation area in the wetland, which also granted recovery of habitat availability and the improvement of the area. The results of the faunistic and hydrobiological element protocol has granted habitat improvement in the area and allowed traditional fish species production with greater gains in weight and size, compared to the conventional closed system.

In general, the results of the project have created sustainable agricultural and fish production and improved the community's life quality, also increased the resilience to flooding and droughts, by means of implementation of practices of the Zenú's intangible cultural heritage. It is important to highlight that incorporation of Zenú elements and strengthening of the community's self-management has granted better appropriation, which also reinforces the commitment with the model's implementation and transformed the area into an educational and tourist attraction in the region.

The adaptation model defined land selection criteria, parameters in hydraulic system and implementation protocol structure designs, which contain traditional knowledge and allows replication in other areas of the ancient Zenú territory and in other areas in the wetland where adaptation techniques prevail (Table 6.5).

Table 6.5 Case study fiche/resume

Location	America, Colombia, Purísima
Duration	72 months
Time Frame	Start date: 01/2013 to End date: 12/2016
Consortium	Corporación Autónoma Regional de los Valles del Sinú y del San Jorge—CVS: Regional Government (environmental authority) Fundación Herencia Ambiental Caribe- FHAC: NGO Asociación de Productores, Pescadores, Agricultores y Artesanos Agroecológicos de Purísima Córdoba - Apropapur: Community organization
Cost	US$285.000
Resources	Human: Community, experts in anthropology, biology, agronomy, hydrology Knowledge: Ancestral practices of the Zenú culture Technologies and Techniques: Agroecological, restoration, faunistic and hydrobiological management protocols
People affected	25 families benefited directly
Donors	Corporación Autónoma Regional de los Valles del Sinú y del San Jorge – CVS, Montería, Colombia: Regional Government - Fundacion Herencia Ambiental Caribe, FAO
Urban Sectors	Safety, Water, Home and living, Tourism and leisure, Construction. regenerative and resilient culture
Resilience dimension	Adaptation
Strengths	– Adaptation practices based on the Zenú culture, against changing ecological systems – Motivation and commitment by the Zenú descending community actors—Knowledge and experience in executing traditional knowledge – Very good coordination among the participating organizations – Excellent results during everyone the years of implementation
Weaknesses	– Project execution made in stages due to budget limitations – Scepticism by some institutions at the beginning – Budget constraints to expose results – It is not possible to finish the assembly of the interpretative museum of the project, due to lack of resources – The existing uncertainty regarding the water pulses of the annual cycle derived from affectations of the upstream basin

(continued)

Table 6.5 (continued)

Location	America, Colombia, Purísima
Impact	This adaptation project has had the following impact on the city of Purísima wetland neighbouring areas: – Generation of sustainable agricultural and fishery production – Life quality improvement of the local community, cultural strengthening and contribution to cultural heritage – Increase of resilience to flooding and drought thanks adaptation constructions based on Zenú culture – Ecotourism and agrotourism strengthening in the city – Wetland preservation by Zenú culture ancestral practices – The Adaptive project managed to demonstrate the effectiveness and efficiency of the pilot and experimental model compared to all other current uses that are being carried out in the swamp and especially those related to extensive livestock
Stakeholder list	– Corporación Autónoma Regional de los Valles del Sinú y del San Jorge—CVS, Montería, Colombia: regional government – Fundación Herencia Ambiental Caribe-FHAC: NGO – Association of Agroecological Producers, Fishermen, Farmers and Artisans of Purísima Córdoba (Apropapur)- Community organization
Challenge	The swampy complex in the lower part of the Sinú river is a Zenú ancestral territory wetland. This complex undergoes important monthly surface area variations and due to the effects of La Niña and El Niño phenomena. The city of Purísima is located next to the wetland and has been impacted by flooding and drought seasons. Management actions used against these natural events comprise dike construction as protection against floods and water pumping in droughts, resulting in hydrological alteration of the wet-land

(continued)

Table 6.5 (continued)

Location	America, Colombia, Purísima
Project description	At the beginning, in 2013, parcels of land were identified in the swampy complex in the Sinú buffer area. The land selection criteria included limnological, sociocultural and infrastructural criteria, soil characteristics and developed agricultural and fish farming activities. The chosen land is the Asociación Apropaur, comprising an area of 6.66 ha located in the wetland north area In 2014, the project was designed according to distinguishing Zenú ele-ments: use of vegetation, wildlife and fish species, architectonic designs and natural resource exploitation. Composed a social, hydraulic, fauna component and a productive component (agroecology and forestry). In the hydraulic component, an open and semi-open system was planned in order to regulate water entry and evacuation according to flow conditions and the wetland's water levels (by contrast with the closed system of fish tanks and watermelon cultivation that was carried out in the area originally). The productive component included wild fishing techniques in canals and reservoirs and traditional agroecological and forestry practices In addition, four protocols were designed: agroecological management, restoration, faunal and hydrobiological management, and finally compost production and management The execution of the project began in the second semester of 2014 and ended at the end of 2019. Sixteen soil modules (ridges), four water reservoirs (open system and semi-open system) were built. The soil modules or ridges are interspersed by water flow channels that determined the hydric pulses from the wetland and define differential dynamics of the system throughout the annual calendar In the crowns of the ridges, wetland sediments were alternately used as well as the aquatic plants from the wetland to put them in the crowns as fertilizer and improve soil moisture. According to the agroecological protocol, food species and other traditional species of fruit trees were planted on the slopes and on the top (crown) of the ridge. The restoration protocol included the planting of native forest species of riparian vegetation in the preservation area. As fauna and hydrobiological management, native fish species that naturally enter the open system were monitored, identifying species, growth, and natural feeding possibilities
Urban resilience el-ements	Urban resilience elements are systems, strategies. Actors, attitudes, heritage and institutional support. Systems include flooding controls and drought management through weather variation and rainy season adjusted systems. Flexible local food production in different weather conditions is also considered, based on traditional exploitation of neighbouring ecosystems of the city. Agents are capable of responding against changes in the ecosystem. Established community actors participate voluntarily in family groups contributing with their piece of land, work and traditional knowledge. Institutions have assisted to the ecosystem's restoration and self-organization mechanisms in order to respond against climate threats

(continued)

Table 6.5 (continued)

Location	America, Colombia, Purísima
List of innovation elements	Innovation in the adaptation project includes: – Rescue and regeneration of the disappeared Zenú tradition and present value – Adaptation to climate variability and rain seasons through an open and semiopen hydraulic system based on Zenú culture practices – Fish/fishery production through the natural insertion of species and nutrients into the canal/reservoir system from the wetland – Agroecology and forest production through the establishment of traditional Zenú species in soil modules (ridges) interspersed by channels with water currents from the wetland, in an open system – Availability of recovery of natural habitat through arrangements of native floristic species and analysis of climatic tensors – Establishment of ecological and historical cultural tourist services, thanks to the adaptation project and the Zenú Interpretative Museum (unfortunately not yet completed due to lack of budget)
Lessons learned	– The identification and evaluation of the lands surrounding the wetland and of the local communities and associative groups allowed the selection of the specific area and helped to implement the adaptation model based on the Zenú culture – The design of an open and semi-open hydraulic system as a prototype of those used by the ancient Zenú culture ensured the potential use of the model's – The design of protocols for agroecological management and composting; restoration; Faunistic management of hydrobiological use granted activity monitoring and defined improvement proposals – The rescue, knowledge and application of the Zenú tangible and intangible cultural heritage and as well as the traditional knowledge inherited from the amphibian culture allowed the recovery of adaptation practices, use of the natural resources of settlers who have been losing their identity, resilience and the heritage of amphibian culture, all of which is detrimental to the protection and care of Cordoba wetlands and the Colombian Caribbean – It can be seen that in the 2014–2019 time horizon, the project's biodiversity indices have increased significantly, to the extent that the flows of water, nutrients, sediments and functional connectivity of the project (intervention area of hydrosedimentological units) with the swamp They have been improving and, in the same way, the arboreal connectivities between roundabouts, dikes, ridges and terraces—which, as indicated in the previous number, has been recovering the level of soil moisture, organic matter and biomass in general. Captures of endemic fish such as the bocachico (in critical threat) are already observed in the reservoirs, as well as birds and reptiles equally threatened in the red books such as: the chavarri, the iguana, the babilla and the turtles, to mention a few examples

(continued)

Table 6.5 (continued)

Location	America, Colombia, Purísima
Upscaling and replication	– This adaptation project may be copied in other places in the city of Purísima and into other wetland neighboring communities in territories inhabited by the Zenú, such as the Sinú, San Jorge, Cauca and Nechí river watersheds. Replication opportunities involve the innovation feature; evidence supported by its efficiency and life quality improvement. Limitations include budget and the local institution's approval – This project's scaling is proposed to wetland neighboring populations affected by floods and droughts in tropical regions, and in territories with communities that maintain traditional practices. Opportunities include improvements in land selection criteria, protocol structuring and design parameters for open and semi-open systems including traditional knowledge. Limitations involve the contributors' competence to expose the project – This project allows perpetuating the biocultural memory of Ancestral knowledge on water management and turns it into a fundamental tool to improve Community Resilience that unfortunately is being lost as the identity of the Zenú Amphibian Culture is blurred by misunderstood development
Web page	This case study does not have its own website, but it is possible to contact Fundaherencia: contacto@herenciaambiental.org
References	[1] Fundación Herencia Ambiental Caribe, & CVS. (2013). Aproximación a la definición de un modelo de adaptación, uso y aprovechamiento de los humedales de Córdoba para la conservación y el manejo sostenible de la Ciénaga Grande del Bajo Sinú. Montería, Colombia [2] Fundación Herencia Ambiental Caribe, & CVS. (2014). Insumos para la formulación del Plan de Adaptación, Uso y Aprovechamiento de los Humedales del Distrito de Manejo Integrado del Complejo Cenagoso del Bajo Sinú. Santa Marta, Colombia [3] Fundación Herencia Ambiental Caribe, & CVS. (2015). Implementación de Protocolos de Restauración Ecológica y Agroecosistémica del Modelo Piloto para la Adaptación, Uso y Conservación del Humedal Ciénaga Grande del Bajo Sinú. Montería, Colombia [4] Fundación Herencia Ambiental Caribe, & CVS. (2016). Implementación de protocolos, investigación aplicada y monitoreo del modelo piloto para la Adaptación, Uso, Restauración y Conservación del Humedal de la Ciénaga Grande del Bajo Sinú. Montería, Colombia [5] Fundacion Herencia Ambiental Caribe- Fundaherencia, & CVS, 2019 "Implementación de un Modelo de Adaptación, Uso, Restauración y Aprovechamiento de los Humedales de Córdoba para la Conservación y el Manejo Sostenible" Fase VI (fase Final). Montería [6] Castaño-Uribe, C., 2020. Informe de Conclusiones Generales y Finales del Proyecto Piloto: Implementación de un Modelo de Adaptación, Uso, Restauración y Aprovechamiento de los Humedales de Córdoba para la Conservación y el Manejo Sostenible 2013–2019. Santa Marta

References

Banco de la República. (2010). *Los Zenúes y el Manejo del Medio Ambiente*. Bogotà D.C.

Carvajal-Cogollo, J. E., Castaño-Mora, O. V., Cárdenas-Arévalo, G., & Urbina-Cardona, J. N. (2007). Reptiles de áreas asociadas a humedales de la planicie del departamento de Córdoba, Colombia. *Caldasia, 29*(2), 427–438.

CVS & Universidad Nacional de Colombia. (2005). *Diagnóstico de la Dinàmica hídrica, tenencia y ocupación del Complejo Lagunar del Bajo Sinú*. Montería, Colombia.

CVS & Universidad Nacional de Colombia. (2007). *Plan de Manejo y Ordenamiendo Ambiental del Complejo Cenagoso del Bajo Sinú*. Montería, Colombia.

De Fontainieu, A. R. (2006). Los camellones, un campo de observación. In F. Valdéz (Ed.), *Agicultura ancestral, camellones y albarradas. Contexto social, usos y retos del pasado y del presente* (pp. 69–70). Quito, Ecuador.

Fundación Herencia Ambiental Caribe & CVS. (2013). *Aproximación a la definición de un modelo de adaptación, uso y aprovechamiento de los humedales de Córdoba para la conservación y el manejo sostenible de la Ciénaga Grande del Bajo Sinú*. Montería, Colombia.

Fundación Herencia Ambiental Caribe & CVS. (2014). *Insumos para la formulación del Plan de Adaptación, Uso y Aprovechamiento de los Humedales del Distrito de Manejo Integrado del Complejo Cenagoso del Bajo Sinú*. Santa Marta, Colombia.

Fundación Herencia Ambiental Caribe & CVS. (2015). *Implementación de Protocolos de Restauración Ecológica y Agroecosistémica del Modelo Piloto para la Adaptación, Uso y Conservación del Humedal Ciénaga Grande del Bajo Sinú*. Montería, Colombia.

Fundación Herencia Ambiental Caribe & CVS. (2016). *Implementación de protocolos, investigación aplicada y monitoreo del modelo piloto para la Adaptación, Uso, Restauración y Conservación del Humedal de la Ciénaga Grande del Bajo Sinú*. Montería, Colombia.

Montejo Gaitán, F., & Rojas-Mora, S. (1992). Asentamientos prehispánicos en el bajo río Sinú y sur de la Serranía de San Jerónimo. *Boletín Museo Del Oro, 34*.

Sánchez Jabba, A. (2013). La economía del Bajo San Jorge. *Documentos de Trabajo Sobre Economía Regional*, (189), 46. Retrieved from http://www.banrep.gov.co/docum/Lectura_finanzas/pdf/dtser_189.pdf

Chapter 7
The Man from the Swamp: Action-Oriented Consciousness, Emotions and Politics

María Alejandra Taborda Caro

Abstract Swamps are traditionally flat ecosystems, with a capacity for seasonal flooding. These bodies of water are connected to rivers, with which they establish a water-land communication relation that serves their daily renewal. These hydric systems work as natural water filters, thanks to their typical vegetation that stores and releases water, and as regulators of river levels. Their shapes are characterized by being horizontally wide and of limited depth. The Colombian Caribbean holds about 82% of the approximately 1900 country's swamps. This region hosts the largest water complex in Colombia and Latin America: *La Mojana*. Following research by the Colombian sociologist Orlando Fals Borda and recent research by the Universidad de Córdoba (UNICORDOBA & Humboldt Agreement, within the framework of the project COL83662-68537, 2017) in this area, this study argues that what has been called 'amphibious culture' and 'swamp men' are still current concepts.

Keywords Amphibian man · Morass · Mojana

7.1 Introduction

By the end of the nineteenth century, from "Geographical Determinism" emerged a way of thinking that established the idea that human, social and political processes are generated by causal effects. According to this view, natural conditions play a decisive role. Against this position, the French historian Lucien Febvre and Vidal de la Blache minted the term "possibilism" to characterize the environmental relationships determined by human actions (Duran, 2016).

At the beginning of this century, the human and the natural emerge as an irreducible duality, which makes evolution and its relationship with nature more "complex" and re-signifies the environmental through the concepts of complexity, post materialism, ecological consumption, convergence and divergence. All this, follows the criticisms

M. A. T. Caro (✉)
Universidad de Córdoba, Córdoba, Colombia
e-mail: mataborda@correo.unicordoba.edu.co

J. Morato et al. (eds.), *Intangible Cultural Heritage and Traditional Knowledge Related to Urban Resilience*, Resilient Cities, https://doi.org/10.1007/978-3-032-12713-6_7

that began with the ideas about the end of nature. Hence polarizing and dichotomous discourses are placed under suspicion. This way of thinking achieved what the environmental decidedly recognizes as "The natural foundations that originate biological processes also develop thinking" (Leff, 2004 p 46).

One of the landscapes where the human and the natural reach a greater rapport in Colombia is *La Mojana*. This sub-region is located in the Colombian Caribbean and has been identified as a strip of highly fertile wetlands, belonging to the Magdalena River basin of Mompox. This subregion has the environmental task of stabilizing the natural channels of the Magdalena, Cauca and San Jorge rivers, cushioning floods and facilitating the decantation and accumulation of their soil sediments (Aguilera, 2004). The wetlands are recognized as playing a mediating role between terrestrial and aquatic systems. They are the transition between a predominantly hydrological and hydromorphic soil with areas of aerated soil where the phreatic level does not reach the surface, covered with shallow waters during most days of the year.

Inhabitants of *La Mojana* are recognized as 'men of the swamps' due to their symbolic practices and their humid daily living. Since the fifties of the last century, the Colombian sociologist Orlando Fals Borda, equipped with qualitative methods and techniques coming from microsociology, such as journals, living with the studied communities for long periods of time, participant observation, recordings, photographs, documents and personal objects provided by the farmers, has identified what he called the amphibian man (Camara, 2017). The following document shows, in three moments, the validity of this denomination in the light of the current experiences of these communities.

7.2 The Origin is not the Beginning

The world of the swamps is dyed by four essentials colours, yellow, blue, brown, and an amalgam of greens in all its shades. In contrast, an ancient smell of humidity constantly and homogeneously lingers in the air and a suffocating heat smother and permeates humanity with scents of grass. The water bodies all have the property of linking the sky and the earth in the distance. In these waters, man is not conceived in isolation but rather closely related to climate, animal cycles, the waters that the river or the spouts bring or carry away, and even the fantastic imagery that is built in the contemplation of the landscape that opens up in front of one's eyes.

The French traveller Luis Striffler was the first to observe and record these characteristics of the amphibian culture in his book "The San Jorge River", which he finished writing in San Marcos working as a public officer in September 1880. The work was published serially in the newspaper "Registro de Bolívar", from issue 212 (June 8, 1886) to issue 275 (August 21, 1886), without continuity. Using the same typographic plates, the newspaper later assembled and published the work as the first edition of a book (Fals Borda, 1986. P. 67).

Nevertheless, much earlier, Sinú mythical thought had already established the amphibious man through the appearance of aquatic or amphibious fauna in the wetlands. It is very telling that its development is chronologically parallel to the development of the hydraulic system that, for Plazas (1981), was configured as a network of waterways that extended approximately 500,000 ha; spouts, irrigation canals and lacustrine terraces to canalize the flow of the rivers, causing, according to the time of year, some areas to be habitable, others flooded for fishing or suitable for agriculture.

There was an intimate connection between the symbolic, the natural, man and his survival. Archaeological excavations, carried out by the mentioned authors, identified remains of snails, catfish, barbet, *hicotea*,[1] *morrocoy*, *babillas*, alligator, brown heron and mammals such as the *guartinaja*, otter, capybara and deer.

If the places mutate, seen against the water, so do the times in their own way. Under these conditions, the natural system manifests its environmental stress in shorter periods of time than in terrestrial environments, allowing the same generation of people to know the swamp in very different states and dynamics. For example, inhabitants of these landscapes can switch among various household types, work practices, or crops in only a few weeks. Human adaptation to those changes has generated a decrease in the possibility of water retention by the wetland. The fertility of the soils has allowed its inhabitants to establish intensive and permanent crops by adapting drainage and land recovery systems, so that when the floods recede, the land is nourished by sediments (Aguilera, 2004) (Fig. 7.1).

Being amphibious requires possessing complete and detailed knowledge of the creatures that share the swamp habitat and establishing a relationship of protection and restoration with them. Their abundance and natural diversity have allowed water men a rich mythological and symbolic configuration that has vastly nourished their culture. In turn, this has resulted in the water men developing diverse representations of their existence, a catharsis that has allowed them to survive, even when only a small number of these species works as symbolic references for the community (Martínez, 2010). For millennia, in these lands or, better said, in these waters, the human condition has been linked to each animal or plant, respecting its special faculty of ferocity, fecundity or love, harbouring all these within the camouflage or knowledge of the climate (Martínez, 2010). From this fusion between the human and the animal thus emerges the *hicotea*-man as a legendary god of the water, to whom language has given the enchantment of converting the biological-animal into the symbolic-animal.

In contrast to the impenetrable entity that Colombia is, which Gómez (2006) defined as a projection of individualism, the amphibious emerges as a collective and political project whose times and origins are not easy to date. It is not solely human, it is not solely natural, its origin is not in the beginning; we can maybe and strangely link it to its permanence.

[1] This an animal from the Colombian Caribbean and Mexico, a kind of turtle that serves as food during Easter. The *hicotea* inhabits swamps and flood-risk areas. It eats vegetation from the swamp. During the dry season, *hicoteas* bury themselfves, living off their body reserves until the new rainy season starts.

Fig. 7.1 The swamp, painting (Granda, 2018)

7.3 There is a Way to Be Amphibious

Water is a life factor that generates possibilities of subsistence for a countless number of inhabitants who have managed to adapt to the plan of water. The water gives a rhythm to all activities and shapes a life style. Inhabitants of *La Mojana* belong to an amphibious culture that has been shaped by the sway of the river floods and their droughts. Their men move confidently by land and water alike, which they have done, as we saw previously, since pre-Columbian times. They make their living by combining agricultural, livestock and wilderness-related activities with river-related activities and fishing. As commented clearly by a river basin fisherman that lives in *La Mojana*: "I've spent a good deal of my life alone and silent, sitting on a boat waiting for the right time, connected to the stars; when fishing is scarce, I also work as a day-paid worker" (UNICORDOBA & Humboldt Agreement within the framework of the project COL83662-68,537, 2017a, 2017b, 2017c, 2017d, 2017e). Thanks to observation of facts such as these, among many other, Orlando Fals Borda got a head start on a theoretical work linked to political action, which systematically spreads and covers a large section of Colombian national history (Moncayo, 2009).

The following are the most significant stories from the inhabitants of some small villages from *La Mojana* such as Ayapel, San Marcos and San Benito Abad, taken from the work "Double History of the Coast", volume III: Resistance in the San Jorge, using the method of argumentative contrast. The stories in this work show the vicissitudes, the character and a way of being of an amphibious person. They have been compared to the experiences of today's inhabitants of these areas to demonstrate

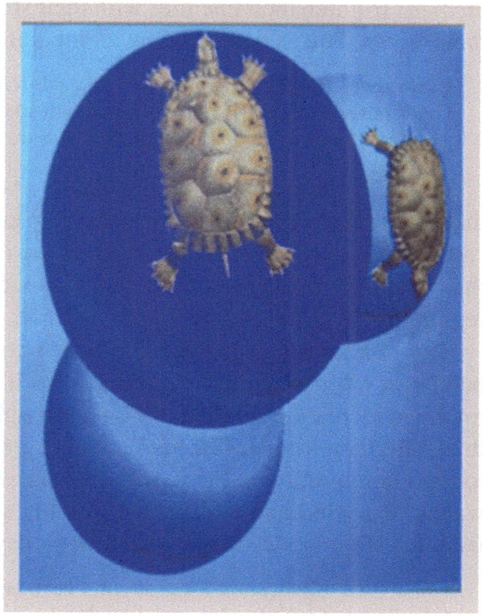

Fig. 7.2 *Hicotea* man, painting, Granda 2018

the permanence of these subjectivities or their mutation. Additionally, the voices of the forgotten authors are included (Fals Borda, 1984) from the same perspective (Fig. 7.2).

The ways of living adopted by the inhabitants of the shores of swamps, streams and rivers adjust to continuous change since they feel imprisoned by the richness that waters bring invariably, by the diversity of fish from inexhaustible puddles, by imperturbable *babillas*, ducks that come and go, *hicoteas* that are born from the muddy soil and reeds used to build temporary huts that the river will take away again.

Some of Fals Borda's profiles of the amphibious person present him/her to be enduring, a good conversationalist, ingenious, and generous. We would also add contemplative but especially adaptive. The man from the wetlands is enduring, with an enormous capacity to overcome the setbacks of life, endowed with a special imagination for overcoming adversity. After a great setback, the man from the swamp resumes his existence with the same energy as before. This harmonizes with what Fals Borda called *sentipensante,* a concept that combines reason and love, body and heart. Men from the swamp identify themselves with the *hicoteas*, with their particular behaviour of inflating their air bladders just as the men from the swamp inflate their lives with the floods, to bury themselves in the mudflats and under the clods in the dry beaches, to endure hunger and thirst.

However, living requires scratching and biting like *the hicoteas* when they grab with their jaws. They do not ever let their prey go, in much the same way as the inhabitants of the swamp do not leave the place where they live when it floods. Other *hicoteas* frighten their enemies with their stinky urine, trying in many ways to defend themselves. This is described by an inhabitant of the riverside village of Pasifueres:

> I have never abandoned my rancha,[2] I fix a different tambo, each time better, with each flood, I have never had electric power, as you will see everything is planned to live like this, there are more floods some years, others less, but we have always borne them.

In his "Double History of the Coast", Volume III Fals Borda (1984:20) clarifies that these people's endurance results from knowing how to wait to satisfy their needs, "although it may seem like fatalism, in the end neither the river men's *rebusque*[3] nor their endurance have the traces of passivity and abulia observed in other peasant regions."

The talk and the pleasant and warm conversation characterize the amphibious being, for when they come into contact with oral myths these populations are prone to mythical recreations of their thinking through their lives. Fals Borda (1986) points out that fishermen from *La Mojana* are also great story tellers, since they must transmit to their community all the wisdom acquired during the long days of listening to the waters, the wind and the birds. Upon his afternoon return from work, a fisherman from *La Mojana*, describes what a work day is:

> I have been fishing since I was seven, I have been doing it since I was a child, I have spent a good part of my life contemplating the sky and the water, in the good as well as in the bad times. Unlike the cultivator, the fisherman is very contemplative, that is the reason why he always looks into the eyes, I have the habit of looking into the eyes of the catfish to imagine its flavour, and in the afternoon after picking up my pay I return to talk and play cards or dominoes.

Nevertheless, the amphibious reaches beyond the unitary genre "man" as male subject. Martínez (2010) shows us that in legends, carnivals, and pre-Hispanic gods there is a duality of sexes or androgyny. The amphibious being moves in the forcefulness of the feminine; the *babilla*, the turtle, the iguana, the snake, all of them figurations that have survived the alligator, in the same plane of collective representation, the maternal is part of a feeling of social cohesion. It is possible to verify that all the community organizations of *La Mojana* have a protagonist that has a female face.

Additionally, Fals Borda (1984) continues by telling us how the amphibious people have invented minor occupations that are important for fishing, cattle farming, cattle herding, seasonal day working and minor trading. They invented water and land *rebusque* as an ingenious, accepted, and productive job. When the drought appears or the flood emerges, people develop enough vision and dynamism to adapt themselves to commercial purposes, "*we have sold everything you can think of: water, egg-stuffed arepas, plantains and even land for embankments*", says Ana, an inhabitant from *La Mojana.*

Hospitality for visitors is another characteristic of the groups who live in the wetlands. The welcoming words, taken from a book of ethnographic notes, of a

[2] A *rancha* is a kind of building whose roof is made of palm.

[3] Activities aimed at making money for survival

social leader to a group of researchers he had just met and who hosted in her house is illustrative of this:

> They can sleep in my children´s beds, this is spiderman's and there is the princess', though we are poor, everything is arranged for your comfort, I will share my family and my poverty with joy", "at the moment there are no fruits" says another woman to some foreigners, "It's been months that we last saw a tree blossom, luckily I have a papaya that I am going to share you now with pleasure.

Finally, creative adaptation is one of the most outstanding characteristics of the amphibious human groups. Even though riverside villagers are aware of the permanent periodicity of the floods and how their traditional way of life has been decomposing and disorganizing, the independent and parcel fisherman accomplishes the transformation of a difficulty into a possibility. Some expressions by the inhabitants of these swamps, recorded through ethnographic description, exemplify the previous points: "*Right now we are flooded but the shoal of fish is good*", replies a fisherman. "*I have improved my technique of making tambos*", announces a lady. "*I found a good campano this year so this wood allowed me to create new improvements in other places* ". "*I have already mastered the tambos, now I have moved forward to make embankments, but this time I mix it with a thicker material. When anyone is on the tambo and the water gets on it, he or she should find a solution.*"[4]

Fals Borda (1984:20A) concludes that people are skilful in everything and can perform equally on land or in water: "The amphibious man of San Jorge and Loba wakes up before the storms come, he does not get upset by the floods, he mocks the snakes and he has successfully developed appropriate technology adapted to their environment."

7.4 There is a Culture of Being Amphibious

The always classic Lévi-Strauss (1972), in his text "Human History and Ecology", affirms up-front that Human History and Natural Ecology appear coupled through the ages as a significant whole, so that there is no separation of the essential relationship between the ecosystem in which a community lives and its cultural creations.

According to Fals Borda (1986:67), amphibious culture is articulated with ideological aspects and psychosocial expressions, attitudes, prejudices, superstitions and legends, which have to do with "rivers, spouts, ravines, hillsides, beaches, swamps and rain forests". This culture also contains community organizations that are empowered to mitigate the impact of ecological structures and to improve the economic base and ways and means of exploitation of natural resources.

The use of this aquatic ecosystem as a receiver of wastewater discharges from neighbouring populations or polluting loads carried by its tributaries from remote areas where they cannot even imagine the existence of wetlands, is negatively

---

[4] Interviews with inhabitants from three villages from *La Mojana*(Ayapel, San Marcos, San Francisco Abadad)

impacting biodiversity and the balance of natural dynamics, producing a growing deterioration which in turn jeopardizes the life quality of human population who live in the swamps.

In addition to what the water brings, as well as what man brings, the cell phone and the news disfigure the imaginary and alter the rhythms and ambitions of the inhabitants of water, who no longer understand or pay attention to it; water is a stranger that threatens them and takes away the dream they want to create (Fig. 7.3).

Nowadays, in the swamps there exist a series of behaviours, beliefs, and practices related to the management of the natural environment and technology, represented in the productive forces and norms of agricultural production, fishing and hunting. These behaviours, beliefs and practices live on the communities and collide with others that have been imposed on them from the outside, stressing the aquatic landscape (Fals Borda, 1986:67). On the impact of the productive forces, an eighty-year-old inhabitant commented:

> The bad fish comes from the hills, where the miners work; from Guarandan people come to buy the fish here because they do not eat the river fish because it is contaminated.

Amphibious culture explains the linear settlement of the hillsides, hamlets and many riverside villages. For example, the concept of amphibious culture helps to understand that the villagers build their houses in a line, on top of dry and narrow ravines that border the fluvial currents where the earth is a limited and valuable material. Therefore, the spatial planning of the resulting lots where the houses are built are not very long because they are met by the swamps or the fences of some large hacienda. The limited nature of the territory allows its inhabitants to walk, ride on a

Fig. 7.3 Encounters with water, painting, Granda 2018

donkey or canoe to their work site every day (Fals Borda, 1986:67). An inhabitant from Cecilia describes her house as it is today:

> As can be observed in Cecilia´s houses, the living room at the entrance, it is small as is the bathroom, the kitchen is the big thing, this [having narrow living spaces] serves to gain a little yard to plant. This house has two bedrooms, in my patio I am almost always visited by the swamp, for this reason it is easier to quarter swamp ducks rather than hens.

Under these conditions, the material objects for daily living, work, food and leisure such as the digger stick, the *machete* and the axe, including complementary elements, configure an amphibious material culture. Fals Borda (1979:43), presents, among others, "*the knife for reaping rice, the hook to get fruits down, the chocó (small seed basket), the grinding stone, the small shovel and the warehouse or elevated deposit. In addition, the fishing and hunting instruments that can be found in most of the riverside houses include canoe, paddle, cast net, hooks, fisto gun, wooden traps, babilla hooks, rubber-made slings*" (p. 67). Thus, there is a range of objects that determine a multiplicity of practices and beliefs adjusted to hunting, fishing and agriculture, herbology and house construction.

In the same way as the majority of houses where amphibious men live communicate with each other, through joint patios that are not enclosed by any fence, to let their owners meet, rivers and water channels become biological corridors that unite the system from within. These corridors constitute the main roads for water and aquatic organisms such as fish that migrate during the initial periods of the floods and for the final phases of the backflow (Aguilera, 2004). Within the ecological system of *La Mojana*, the imaginary vision that people have of their environment determines how they relate to it. The main vegetation coverage is composed of natural forests, pastures, stubbles, and aquatic vegetation typical of swamps, shallows and shores. These last ecosystems serve as important shelter and protection of the coastal fauna, supply of surpluses for the aquatic habitat, and a supply of food (Aguilera, 2004:20). In relation to the impressions that the community has about its environment, an eighty-year-old lady answers some questions like this:

> Do you guys here eat any kind of capybara, hicotea? mm. There is not any capybara, in case there is one, it could be found around those bad zapales. Is the zapal bad for you? Yes … zapal is bad, what I am telling you is that on firm land, nothing walks by other than the tiger. The tiger (laughs). Impenetrable for people, the zapales. People do not walk around there, around that firm land.

A preliminary conclusion to be drawn is that there still remain and survive various conducts and behaviours that typify *Mojana* inhabitants as belonging to an amphibious culture, configured by a set of beliefs and practices that match the management of the environment; the human groups that live there have come up with their own occupations, basic strategies of reproduction, adaptation, and environmental mitigation. They still face the hardness of collective defence practices, although it is necessary to review this last aspect in more detail, for it seems to have been modified in each of the villages and towns along the rivers. The amphibious culture is still composed of the elements observed by Fals Borda, ideological in nature, that articulate psychosocial expressions, attitudes, prejudices, superstitions and legends that have to do with rivers; all of the above as a vital substratum for the total modification of the worldview of wetland people, a process that slowly turns old beliefs worthless and that plants the seed of a radical change in the face of life.

References

Aguilera, M. (2004). *La Mojana: riqueza natural y potencial económico.* Retrieved from http://www.banrep.gov.co/sites/default/files/publicaciones/archivos/DTSER-48.pdf, http://www.bdigital.unal.edu.co/1401/4/01PREL01.pd

Camara, D. (2017). *Constribuciones del pensamiento social latinoamericano para los estudios organizacionales: Fals Borda, Marini y Fernández Retamar.* Brasil: Administración y Desarrollo.

Convenio UNICORDOBA & Humboldt, en el marco del proyecto COL83662-68537. (2017a). *Linea de tiempo vereda Pasifueres, municipio de San Marcos, departamento de Sucre.* Montería.

Convenio UNICORDOBA & Humboldt, en el marco del proyecto COL83662-68537. (2017b). *Notas de campo de la comunidad de Cecilia, municipio de Ayapel, Córdoba.* Municipio de Ayapel.

Convenio UNICORDOBA & Humboldt, en el marco del proyecto COL83662-68537. (2017c). *Notas de campo de la Mojana de Córdoba y Sucre.* Municipio de San Marcos.

Convenio UNICORDOBA & Humboldt, en el marco del proyecto COL83662-68537. (2017d). *Notas de campo, a partir de observaciocion participante con la comunidad de la vereda de Cuenca, municipio de San Marcos, Sucre.* San Marcos.

Convenio UNICORDOBA & Humboldt, en el marco del proyecto COL83662-68537. (2017e). *Notas de campo, observación participante con la comunidad de Cecilia del municipio de Ayapel, Córdoba.* Municipio de Ayapel.

Duran, J. (2016). *"DETERMINISMO Y POSIBILISMO" GEOGRAFICO.* Retrieved from http://profjoseduran.blogspot.com.co/2016/11/determinismo-y-posibilismo-geografico.html

Fals Borda, O. (1979). *Historia doble de la Costa 1 MOMPOX Y LOBA.* Retrieved from http://es.calameo.com/read/001456652c149ec445c4a

Fals Borda, O. (1984). *Historia doble de la Costa 3. RESISTENCIA EN EL SAN JORGE.* R

Fals Borda, O. (1986). *Historia doble de la Costa, 4 Vols. Bogotá: Carlos Valencia Editores 4 Vols.* Bogotá: Carlos Valencia Editores.

Gómez, J. G. (2006). *Colombia es una cosa impenetrable.* Bogotá.: Diente de León.

Leff, E. (2004). *RACIONALIDAD AMBIENTAL La reapropiación social de la naturaleza.* Retrieved from https://bibliodarq.files.wordpress.com/2014/12/leff-e-saber-ambiental-sustentabilidad-racionalidad-complejidad-poder.pdf

Lévi-Strauss, C. (1972). *Estructuralismo y Ecología.* Anagrama.

Martínez, C. (2010). *El ser anfibio. Imaginarios acuáticos en el Sinú. Análisis de la fauna mitológica en la tradición oral y la orfebrería Sinú.* Cali. http://www.univalle.edu.co/~estudiosliterarios/imagenes/pdf%20tejedora%20modificados/EL%20SER%20ANFIBIO.pdf

Moncayo, V. (2009). *Fals Borda: hombre hicotea y sentipensante.* Retrieved from iblioteca.clacso.edu.ar/clacso/se/20160304050141/01pres.pdf

Plazas, C. y. (1981). *Asentamientos prehispánicos en el bajo río San Jorge.* Bogotá: Fundación de Investigaciones Arqueológicas Nacionales. Banco de la República, Bogotá.

Chapter 8
Appropriate Technologies for Building Resilient Communities

Jordi Morató, Luis Diaz, Carlos Arias, Nicola Tollin, Alberto Inocenti, Amado Enrique Navarro Frómeta, José Luis Marrugo Negrete, Josep María Bayona Termens, Jorge Antonio Herrera Cárdenas, Lorenzo Morales Morales, Jorge Montoya Restrepo, Joe Sánchez Marín, José Gregorio Ortega, José José Palacín Salcedo, and Germán Holland Enamorado Sánchez

Abstract Climate change, rapid urbanisation and the progressive ageing of centralised infrastructures are placing unprecedented pressure on water supply and sanitation, particularly in the small rural communities and vulnerable territories of the Global South. This chapter argues that building community resilience requires

---

J. Morató (✉)
UNESCO Chair on Sustainability – Polytechnic University of Catalonia, Barcelona, Spain
e-mail: jordi.morato@upc.edu

L. Diaz
Polytechnic University of Catalonia, Barcelona, Spain
e-mail: luis.david.diaz@upc.edu

C. Arias
Aarhus University, Aarhus, Denmark
e-mail: carlos.arias@bio.au.dk

N. Tollin · A. Inocenti
University of Southern Denmark, Odense, Denmark
e-mail: nto@iti.sdu.dk

A. Inocenti
e-mail: alinn@iti.sdu.dk

A. E. N. Frómeta · J. A. H. Cárdenas · L. M. Morales
Universidad Tecnológica de Izúcar de Matamoros, Matamoros, México

J. L. M. Negrete · J. G. Ortega · G. H. E. Sánchez
Universidad de Córdoba, Córdoba, Colombia

J. M. B. Termens
Instituto de Diagnóstico Ambiental y Estudios del Agua, Barcelona, España
e-mail: jbtqam@cid.csic.es

J. M. Restrepo
Tecnológico de Antioquía, Medellín, Colombia
e-mail: jmontoya@tdea.edu.co

J. Morato et al. (eds.), *Intangible Cultural Heritage and Traditional Knowledge Related to Urban Resilience*, Resilient Cities, https://doi.org/10.1007/978-3-032-12713-6_8

a shift from conventional, capital-intensive centralised systems towards appropriate, decentralised and nature-based technologies adapted to each local social, economic and environmental context. Drawing on the Water Sensitive City framework and the concept of Nature-Based Solutions (NBS), the authors examine treatment wetlands (constructed wetlands) as a low-cost, low-energy and replicable option for wastewater treatment and water-quality restoration. The performance of pilot systems implemented in Izúcar de Matamoros (Mexico) and Montería (Colombia) is evaluated, assessing the removal of conventional pollutants such as BOD5 and of emerging contaminants such as the fragrance Galaxolide and the pesticide cypermethrin through physical–chemical and GC/MS analyses. The results confirm that constructed wetlands achieve high treatment efficiencies under tropical conditions, offering clear advantages over conventional plants. The chapter concludes that the main barriers to wider adoption in Latin America are not technical but social, economic and political: lasting success depends on community ownership of the technology and on external actors approaching each local context with humility and respect for its social, ethnic and geographical specificities.

Keywords Appropriate technology · Nature-based solutions · Constructed wetlands · Decentralised water treatment · Community resilience · Emerging contaminants

8.1 Introduction

Developed countries face a troubling combination of aging centralized water treatment and delivery systems and growing water scarcity and contaminant issues. Many countries lack the public funding necessary to address looming infrastructure issues. In the United States, the Environmental Protection Agency (EPA) anticipates that smaller, centralized water treatment systems may have difficulty meeting the rigorous standards imposed by the Water Safety Act.

Developing countries have to innovate and implement new solutions to fill the existing infrastructure gap regarding urban water management (Noyola et al., 2012). Therefore, **the future of technology to be implemented for urban water management for the countries lacking infrastructure should shift towards solutions, that are resource oriented, integrated, sustainable, distributed, and nature based** (Masi et al., 2018).

A new **socio-technical resiliency, which can mitigate the uncertainty of climate change hazards** it should be defined. New concepts such as Water Sensitive Cities, Water Sensitive Urban Design or Sponge Cities have been appearing in recent years.

J. S. Marín
Municipality of Medellín, Medellín, Colombia

J. J. P. Salcedo
Universidad del Atlántico, Barranquilla, Colombia

8.2 Water Sensitive City

The concept of **Water Sensitive City (WSC)** has been introduced, as stated goal, by Australian Commonwealth's National Water Initiative (Brown et al., 2009a) under the label: "Innovation and Capacity Building to Create Water Sensitive Australian Cities". The Australian Government was trying to introduce a new water management approach which integrated social and institutional capital with diverse and sustainable technology to protect and enhance water in urban environment. The need of the Australian government to change policies on water comes from the rising awareness about climate change and water-wide disasters resulting from it (Wong & Brown, 2009b).

Hence, one of the main topics and challenges in reaching the idea of WSC is **seeking a socio-technical resiliency, which can mitigate the uncertainty of climate change hazards**. Indeed, an urban resilient system is capable of harnessing external disturbances to create new opportunities for innovation and development. The ability to change under the pressure of external events needs a socio-technical perspective that combines human governance and complex infrastructure to develop a new perspective (Folke, 2006).

The Water Sensitive City vision is not an absolute principle but is presented in the literature as a set of factors specific to the context of analysis that led a city **to be more sustainable in water management**. To reach a more sustainable approach in urban water management, it is common to talk about the 'hydro-social contract' (Lundqvist et al., 2001; Brown et al., 2009b, 848) as an implicit agreement between communities, governments, and business on the governance of water. The hydro-social contract aims to find a proper way to "transform cities through reconnecting best thinking and practice in urban water management, urban design, and social and institutional system" (Wong & Brown, 2009a, 675).

Brown et al. (2009a) have defined an urban water transition theory which analyses the historical, ecological, demographic, and climatic evolution of cities through the concept of the hydro-social contract. The theory, defined as **Water Urban Transition Framework**, has been developed as a benchmarking tool to help local governments in transforming cities to more water sensitivity and sustainable environment (Buurman & Padawangi, 2018). According to Brown, to reach the state of WSC it is necessary a transition through six different city-states characterized by a specific hydro-social contract modelled on a technological, ideological, and institutional background. The transition framework is a sequence of phases throughout rigid and conventional systems to more flexible and diversified urban states: Water supply city, Sewered city, Drained city, Waterways city, Water cycle city and Water sensitive city (Brown et al., 2009b).

Each state is characterised by two specific indicators: the "Cumulative Socio-Political Drivers", which reflect the evolution in demands and expectation about water management; and the "Service Delivery Functions", which represent the translation of Drivers in concrete services (Brown et al., 2009a). Through the analyses of the evolution of these indexes, it is possible to define the hydro-social contract of each state. The contract is structured as it provides diverse and flexible technologies, infrastructure, and urban forms, designed to reinforce sustainability and social capital within a strict collaboration between society and the technology sector (Wong & Brown, 2009b).

The state is combined in a nested continuum whereby the skeleton of the later city state forms the basis of the following state (Brown et al., 2009b). The transition model has been represented as a linear path, but this does not preclude the possibility of leapfrogging from not consecutive states (Barron et al., 2017). Indeed, the purpose of the water management transition framework is to represent a set of water infrastructure related to specific institutional and community attitudes that reveal the city journey towards a water sensitivity (De Haan et al., 2015).

As developed in Brown framework, the six city-states are defined as (Brown et al., 2009b):

Water Supply City. The basic state of modern water management characterised by a centralised system that ensures the supply and distribution of water economically and fairly using infrastructures like pipes and dams. The community believes that water is an unlimited and harmless good for the environment.

Sewered City. A sewered city aims to ensure public health protection devising a new system for collecting and removing wastewater to reduce the proliferation of diseases and infections. Hence, a sewerage system is built to deliver wastewater to a receiving waterway. As in the earlier state, discharging wastewater into natural receiving waterways is not considered as an environmental problem to the community.

Drained City. A drained city aims to protect urban environment from flooding, caused by the rapid process of urbanisation and natural soil sealing. Consequently, the institutions involve new urban water service providers and policy for stormwater management through the implementation of drainage and channelisation dumping ground systems to divert water to receiving waterways.

Waterways City. The state of waterways city changes completely its perspective. The environmental impacts on water embracing the necessity of new approaches. The rise of pollution of waterways and the social and aesthetic needs of green open space in the urban environment, change the institution perspective towards water urban management. Community and institutions attention are addressed to new decentralized and green infrastructure which provide filtering of polluted waterways and new green recreative urban solution for social health and amenity.

Water Cycle City. The water cycle city state derives from the recognition of the limits of natural water resources. Hence, stakeholders consider the possibility of a new integrated urban water management focused on new practices of water conservation and unconventional water supply solutions. The water cycle city needs a co-management approach between business, communities, and the government to find new water conservative and protection policies.

Water Sensitive City. The water sensitive city state is framed on the idea of integrating water cycle in urban management to reach a wide of benefits for a better liveability and resilient to the new climate challenges. Fundamental is the strict collaboration between society and technology: community is actively engaged and sensitized to water in the city; innovation and technology in planning have to create adaptive and multifunctional infrastructure to protect the health of waterways, mitigate flood risk and create new opportunities through the design of green public spaces.

The six stages described by Brown can be organized in two different macro group: the supply, sewerage and drainage cities against waterways, water cycle and water sensitive cities. The first group cannot be considered as a real transition (change) because the political and administrative regimes is basically unchanged, but each state add more utilities to the existing portfolio. Contrariwise, the last states deeply modify the establishment of existing infrastructure and policies in order to overcome new challenges related to the climate and environment (De Haan et al., 2015).

Reflecting the water management transition theory, three pillars have been recognized as fundamental to translate the water sensitive city's hydro-social contract and system resiliency structure into **fundamental principles for practice** (Wong & Brown, 2009b):

(a) **Cities as Water Supply Catchment**. The rising pressure on urban water supplies has increased the awareness about the alternative solution to collect and use water a part of conventional infrastructure. It becomes necessary to access to a diversity of water sources via centralized and decentralized infrastructure through adaptive and multi-functional infrastructure (Lloyd et al., 2012). The diversity of sources is associated with harvesting, treatment, storage, and deliver water in urban environment to minimizing the import and export of water from outside the city boundaries. Cities have access to water inside the dynamics of the urban water cycle composed of potable water, stormwater, and wastewater systems. Conventional approaches in urban water managing use principally potable water and wastewater, while stormwater is not seen as a potential water source yet (Dobre, 2013).

To transform cities into water supply catchment, it is necessary to replace or implement conventional infrastructures with decentralised infrastructures capable of reducing the impact on fresh water, managing the water resource locally. Using decentralized infrastructure provides diversify in local water utilities with the positive consequence of reducing fresh water demand and usage; upgrading aged and obsolete technologies; delivering more adaptive systems to adverse event (Leigh & Lee, 2019).

(b) **Cities Providing Ecosystem Services**. The rise of environmental problems and pressure on urban water systems provides the new idea of ecosystem services for built and natural environments in cities, trying to reach harmony between water planning and urban planning (Lloyd et al., 2012). Cities need to become more sustainable and resilient to climate change, global warming, and the impact of urbanization on the natural ecosystem. In general, ecosystem services related to urban planning comprise the design of public spaces as a source of amenity for the wide-city concept (urban space and citizen). The complexity of this point lies in the **definition of practical solutions capable of conferring functional but environmentally friendly spaces**. To achieve the goal, it has stated the idea of **implementing blue and green infrastructure related to the concept of WSUD especially in stormwater management, as quality and quantity**. WSUD largely contributes to the idea of rehabilitating waterways through the union of treatment technologies and urban design solution **using Nature-Based Solutions** such as wetlands, bioretention system (defined below).

(c) **Cities Comprising Water Sensitive Communities**. The concept of Water Sensitive City involves a whole series of principles related to the technical and the social sphere of urban water management. For new water management technologies to be effective, they must be socially related and accepted by local institutions to ensure their successful implementation. Furthermore, to reach a Water sensitive community, the social socio-political contribution to sustainability and water sensitive behaviour needs a strict collaboration between science, policy, institutions, and community (Lloyd et al., 2012). The importance of creating a benchmarking network between public and private institutions at a different urban scale, encourage the inclusion of water-related issues in the political administration priorities. In conclusion, the implementation of decentralised infrastructures allows the adoption of water management solutions on a different scale. This multi-scale capability is essential to ensure that the community is brought closer to water, as it provides the opportunity to involve citizens in planning processes or awareness-raising campaigns about water and the environment in general (Leigh & Lee, 2019).

Although there are many works in the literature about the concept of Water Sensitive City, it seems to be an ideal concept because it does not exist any tangible examples of a water sensitive city yet. Reports can be found about some cities (especially in Australia) which are going through achieving the state of water sensitivity, but nowadays, there are still too many barriers and limits to the concretization of the concept (Brown et al., 2009a; Ferguson et al., 2013a, 2013b; Rijke et al., 2012; Wong & Brown, 2009a).

It can be stated that there are no fundamental shortcomings related to technological advancement and urban design solutions, rather there is a clear difficulty in the implementation and dissemination of these solutions in urban planning and management. Literature on water sensitive city and sustainable urban water management agree that the gap between theory and practice related to the achievement of what is defined as the Water Sensitive City stage through the review of conventional approaches is largely caused by (Ferguson et al., 2013a, 2013b): socio-institutional barriers, lack of proper community involvement and awareness, lack of benchmarking tools and sharing between states for the dissemination of knowledge and practices, lack of tools for the analysis of economic and ecosystem benefits derived from the implementation of innovative technologies (Ferguson et al., 2013a, 2013b; Floyd et al., 2014; Rijke et al., 2012).

As for the **inadequacy of the government machine, still strongly linked to conventional methods and solutions, it is ineffective because it is called to face problems characterized by factors such as uncertainty, complexity and variability** (Floyd et al., 2014). For this reason, it is **necessary to change approach and identify an intervention strategy capable of increasing decision-makers' confidence in the use of new solutions**, even if risky (Dobbie et al., 2016).

Some research (mainly conducted in Australian cities) has shown how the implementation of strategic plans, planning scenarios and workshops, involving (Rijke et al., 2012) state and local governments, water utilities, private companies, planners and homeowners, allow to share knowledge and establish collaborative relationships fundamental to innovate the urban water management approach (Ferguson et al., 2013a, 2013b). Within the water management sector there are several organisations, all with a specific function and interests. Through the introduction of collaboration, co-management and government partnership mechanisms, it is possible to create an environment of trust and collaboration between decision-makers and decision-makers in order to build practical experience of long-term, flexible and adaptive solutions (Dobbie et al., 2016). Lastly, the inclusion of homeowners within the dynamics of urban water management allows to educate and raise awareness in the more sustainable management of water at batch scale.

The last element, highlighted as an obstacle to the implementation of the WSC concept, is the **absence of an interdisciplinary approach that does not allow the identification of flexible strategies capable of convey with complexity** (Farrelly & Brown, 2011). The transition to the concept of water sensitive city is defined as a "wicked problem" (Floyd et al., 2014), as any type of (conventional) solution that is applied resists solving the problem. For this reason, it is necessary to break down the disciplinary barriers erected by the evolution of a silo mentality that has always seen urban water management as an exclusive engineering competence (Rijke et al., 2012).

According to Barron et al. (2017), to **break down disciplinary barriers is to integrate social sciences with engineering sciences through the implementation and collaboration of 3 thematic groups corresponding to WSC principles: technology and innovation, urban planning and design, governance, and society**.

The outcomes from the collaboration of these disciplinary fields should provide knowledge and awareness of the actors involved in water management. In this way it is possible to develop technologies and tools that respect the principles and guidelines defined by stakeholders and social needs and that can be easily implemented in urban planning and design practices (Barron et al., 2017).

By reporting what are the barriers, present in the scientific literature, to the implementation of the new WSC paradigm, it is possible to trace the issue back to a general rigidity consolidated in urban water management, partly due to the lack of adequate knowledge among planners regarding the understanding of the mechanisms of the water cycle (Rijke et al., 2012). Resuming the concept of interdisciplinary sharing and collaboration, Serrao-Neumann et al. propose the use of other principles related to water management and administration to fill the gap in knowledge needs of water sensitive city planners. One of these principles is that of Urban Water Metabolism, whose evaluation of the use and consumption of water resources in urban contexts provides numerous food for thought regarding the efficiency of water supply tools, greater knowledge of water flow processes in urban contexts, tools for the control of pollutant discharge in waterways and tools for economic-financial analysis related to the implementation of innovative solutions related to water system management.

The lack in knowledge can also be filled by introducing benchmarking tools capable of defining a right approach to lead cities through water sensitivity reducing ambiguity and giving an endorsed and replicable tool to evaluate how policies, strategies and solutions are effective within the water-sensitive approach (Rogers et al., 2020).

Recently, an in-depth study of the indices related to water system management has been carried out and an all-inclusive index has been designed to jointly analyse the range of social, governance, economic, multifunctional, and adaptive attributes. The index, defined **WSC Index**, is composed of 7 goals for a total of 34 indicators that allow to evaluate different aspects of a city and provide a general vision of how the policies and solutions adopted suite of WSC objectives. The WSC Index has been tested on Australian realities but other research sees its implementation also in different contexts and scales (Özerol et al., 2020). It should be noted that the index is a tool for benchmarking and evaluating a city's progress towards transition to a water-sensitive city but neither certifies nor provides practical solutions to overcome issues.

Finally, it should be noted that no direct reference to economic or financial issues regarding the WSC concept has been found in the literature. Only in a few cases was found the absence of a tool capable of assessing the cost–benefit of implementing water-sensitive solutions or the difficult replicability of these solutions to different geographical and political contexts (Lloyd et al., 2012).

There are many examples in the literature where the definition of the term WSC appears in association with that of Water Sensitive Urban Design (WSUD). The concept of Water Sensitive Urban Design appeared for the first time in 1992 in a publication by Mouritz (Fletcher et al., 2015) in which the design of urban water management spaces had to aim at the balance between water supplies and water hazards; the maintenance of water quality and the protection and conservation of natural spaces by providing water-related recreational opportunities.

In the 2000s the concept began to evolve, and scientific opinion began to recognise in the principles of WSUD those of integrated management of the urban water cycle (i.e. clear water, stormwater, and wastewater) (Hoyer et al., 2011). In this way, the evolution of WSUD principles began to be confused with those of WSC theory even though the latter is the stage of a city where urban water management is planned through the implementation of WSUD-related techniques and practices (Fletcher et al., 2015). In fact, the same meaning of the term highlights the need to bring water sensitivity into cities landscape, through urban design, combining traditional planning approach with environmental and engineering science (Ashley et al., 2013). In this way water becomes an element of urban planning which must consider solutions not only related to the maintenance of water resources and the environment but also to provide protection from damage related to climate events and provide a series of related environmental benefits on multiscale (Dolman et al., 2013). According to the Royal Haskoning agency, there are some indicators and parameter to follow in order to apply WSUD principles to transform future cities in water sensitive cities (Dolman et al., 2013):

- liveability of urban environment
- sustainable use of water resources
- climate control solutions
- energy from water
- protecting water quality
- water-resilient spatial solutions

In conclusion, the approach according to WSUD has developed over the years numerous practical-technological solutions closely related to urban and architectural planning. These technologies mainly concern (Hoyer et al., 2011) systems:

(a) rainwater harvesting
(b) bioretention
(c) biotopes
(d) retention system
(e) infiltration zones
(f) swales
(g) detention ponds
(h) drainage surfaces.

The term **Sponge City** has actually gained a huge amount of support, especially in China and is another term on the growing list next to regenerative, sustainable, green, eco, resilient, low-impact, future proofing, zero-carbon, and the list goes on. The *Sponge City* indicates a particular type of city that does not act like an impermeable system not allowing any water to filter through the ground, but, more like a sponge, actually absorbs the rain water, which is then naturally filtered by the soil and allowed to reach into the urban aquifers. This allows for the extraction of water from the ground through urban or peri-urban wells. This water can be easily treated and used for the city water supply. Sponge cities are therefore, cities structured and designed to capture rain water and utilize it to reduce floods.

The **"Sponge City" initiative is a Chinese national policy framework that focuses on sustainable urban stormwater management in 16 Chinese cities**. The project allocated to each city between 400 and 600 million yuan for the implementation of innovative water management strategies that would gradually transform these cities into "Sponge Cities". In 2016, Shenzhen became a pilot city of the "Sponge City" initiative and has been working to help urban communities in Shenzhen become more resilient to urban flooding with green infrastructure. A massive project, the Sponge Cities Initiative is a collaborative effort among city regulators, property owners, and engineers. The cities of Xiamen and Wuhan have already produced effective Sponge City projects. While the initiative faces some constraints related to funding and enforcement, its initiative has an ambitious goal: by 2020, 80% of urban areas should absorb and re-use at least 70% of rainwater.

Sponge Cities utilize various techniques green infrastructure retrofits such as permeable pavements, green rooftops, bioswales and constructed wetlands at a variety of scales to reduce the intensity of rainwater runoff by enhancing and distributing absorption capacities more evenly across targeted areas. The resulting groundwater replenishment increases availability of water for various uses and serves to not only reduces flooding but also enhances water supply security. **A sponge city needs to be abundant with spaces that allow water to seep through them. Instead of only impermeable concrete and asphalt, the city needs more:**

- **Contiguous open green spaces,** interconnected waterways, channels and ponds across neighbourhoods that can naturally detain and filter water as well as foster urban ecosystems, boost bio-diversity and create cultural and recreational opportunities.
- **Green roofs** that can retain rainwater and naturally filters it before it is recycled or released into the ground.
- **Porous design** interventions across the city, including construction of bio-swales and bio-retention systems to detain run-off and allow for groundwater infiltration; porous roads and pavements that can safely accommodate car and pedestrian traffic while allowing water to be absorbed, permeate and recharge groundwater; drainage systems that allow trickling of water into the ground or that direct storm water run-off into green spaces for natural absorption
- **Water savings and recycling**, including extending water recycling particularly of grey water at the building block level, incentivizing consumers to save water through increased tariffs for increase in consumption, raising awareness campaigns, and improved smart monitoring systems to identify leakages and inefficient use of water.

Sponge cities are designed to tackle with the following 4 challenges:

Less water available in urban and peri-urban areas. Extensive urbanization and urban sprawling led to the formation of thousands of square kilometres of impermeable areas made up of impervious roads, pavements, roofs and parking lots that do not allow water to be absorbed into the ground but that simply collect the rainwater through the urban drainage infrastructure and channel it into rivers, lakes or into the sea. This traditional type of design led to the creation of cities which are increasingly impermeable and have an increasingly greater impact on the natural water cycle. In practise this means that since less rain water is allowed to filter through the urban soil, less water is available to be extracted from aquifers in urban and peri-urban areas.

Polluted water discharged into rivers or the sea. Another key issue is related to the fact that rain water and wastewater (namely water from our sinks and toilets) is collected by one single drainage system. This drainage system (imagine one big pipe) collects all the rain water (when it rains) and the wastewater from our houses and directs it to a wastewater treatment plant where it gets treated before it is discharged again into rivers or the sea. When it rains, many times the wastewater treatment plant cannot accommodate all the water that the drainage systems carry. Therefore, much of the rain water mixed with the wastewater is discharged untreated into rivers. The more impermeable the city is, the more water will be mixed with wastewater and will not be able to be treated but discharged directly into rivers. This increases the level of pollution of local water bodies.

Degradation of urban ecosystems and green areas due to sprawling. This led to a considerable loss of urban biodiversity, a drop in available green areas for natural ground filtration of storm water, a decrease in CO_2 capture by plants, fewer spaces for natural cooling through urban green microclimates and generally less liveable, healthy, comfortable and attractive public spaces.

Increase in the intensity and frequency of urban flooding. particularly considering predicted increase in extreme weather events due to climate change. As the absorbing capacity of the urban surface is decreased, storm flooding risk is increased. Flooding leads to increased groundwater pollution and has considerable impact in terms of damage to properties and health related issues.

There is **wide range of benefits associated with the implementation of sponge cities** (Bao-Jie He et al., 2019; Jenkins, 2020). These include:

- More clean water for the city. Replenished groundwater and thus greater accessibility to water resources for cities. This also entails greater water self-sufficiency which allows cities to increasingly rely on water sources from within their boundaries
- Cleaner groundwater due to the increase volume of naturally filtered storm water. This means lower environmental and health costs due to considerable decrease in water pollution
- Reduction in flood risk as the city offers more permeable spaces for the natural retention and percolation of water. This leads to better resilience and in particular greater ability to deal with higher flood risks resulting from climate change
- Lower burdens on drainage systems, water treatment plant, artificial channels and natural streams. This also entails lower costs for drainage and treatment infrastructure
- Greener, healthier, more enjoyable urban spaces. Greener urban spaces improve quality of life, create more pleasant landscape aesthetics and recreational areas that are enjoyable and attract people. This also means increase in land value due to aesthetically more pleasing, cleaner and healthier open spaces close to private properties
- Enriched biodiversity around green open spaces, wetlands, urban gardens and green rooftops.

8.3 Decentralized Water Treatment

As the climate changes and water supplies are depleted by municipalities, agriculture, and industry, there's a growing need for effective and efficient treatment in remote locations around the world. In addition, the energy to power these plants and the chemicals needed for treatment create environmental and financial impacts that must be minimized.

As we stated before in Chapter 4.1.1, water scarcity is one of the greatest challenges of the twenty-first century, affecting not only human health and the environment, but also local economies, which depend on water for business, industry, and agriculture. Theoretically, there is more than enough fresh water on Earth to meet human needs. However, it is unevenly distributed, and much of it is not easily accessible, including water stored in snow and ice, remote lakes and rivers, and deep underground.

On the other hand, salt and brackish water are far more plentiful than fresh water, yet without treatment to remove the salt, they are unusable. Meanwhile, much readily obtainable fresh water has been polluted by agricultural, municipal, and industrial sources—as well as literally flushed down the toilet.

These facts add up to an **increasing need for sustainable water and wastewater treatment facilities, but conventional designing, financing, permitting, building, and commissioning of new plants may take years and the costs can be huge**.

Decentralized treatment is simply the practice of locating water and wastewater treatment plants at the site of water supply, demand, or ideally both. It's a flexible, sustainable alternative to large treatment plants that require miles of expensive supply and delivery infrastructure.

Decentralized treatment is an important solution for global water scarcity and climate resilience challenges for a number of reasons:

- Locating plants near the source of supply and need eliminates the prohibitive costs of pipelining to distant plants.
- Scalable and portable plants avoid the prohibitive expense of large projects and limit the risk of economically crippling "white elephant" plants that can be obsolete before they're even paid for.
- Lower energy requirements enable off-grid operation on alternative energy sources.
- Smaller-scale projects may eliminate politically challenging negotiations across diverse regions and facilitate local control of community water treatment systems.
- In industry, decentralized treatment can provide quality control for individual sites and improve community relationships by reducing ecological impacts and reliance on local water resources.

A decentralized water and wastewater treatment strategy can provide the efficiency and flexibility needed to deal with water challenges around the globe.

Although the advantages of decentralized treatment have long been recognized, technological and logistical challenges have limited its application. But new advances in both technology and packaging have widened decentralization's potential spread. However, most of the advantages is the extensive use of Nature-Based Solutions.

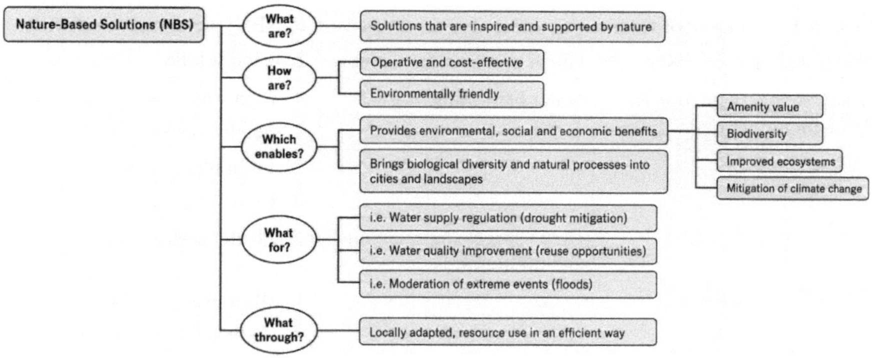

Fig. 8.1 Conceptual framework of NBS solutions for water management in the urban context (Data from Faivre et al., 2017; O'Hogain & McCarton, 2018)

8.4 Nature-Based Solutions (NBS) for Adaptive Design

As we stated before in the 4.1.4 section (Natural Treatment Systems), developing countries are in a need of better and more environmentally sound urban water management that can take advantage of the country's climatic conditions and that offers a good investment solution, because the capital for infrastructure in these countries is usually scarce (Arias & Brown, 2009).

Nature-Based Solutions (NBS) are water infrastructure alternatives inspired by these principles. NBS can be defined as technical solutions that mimic and are supported by natural processes, which are cost-effective, provide environmental, social and economic benefits and help to build resilience (Faivre et al., 2017). Figure 8.1 shows the conceptual framework that integrates the aspects involved in NBS concept.

Taking into account the conceptual framework, Table 8.1 shows at least ten solutions that can be considered as part of NBS projects for improving urban water management in developing countries. These NBS examples refer to the natural or seminatural systems that provide options with benefits that are equivalent or similar to conventional gray water infrastructure (Bernhardt et al., 2018).

8.5 Local Action Tools: Treatment Wetlands and Water Quality Restoration

The foundations of water management are affected by the uncertainty levels related to that resource. These levels are varying as a result of changes in demographic trends, consumption patterns, migration and climate change, resulting in increased pressure on water resources. Adapting to these uncertainties and to develop strategies to

Table 8.1 Examples of solutions considered in the framework of NBS (Bernhard et al., 2018)

Water management issue	Green infrastructure solution	Specific solution
Water supply regulation (i.e. Drought mitigation)	Water harvesting	1. Rain water harvesting systems
		2. Fog harvesting systems
		3. Infiltration pools
		4. Rain Gardens
	Green spaces (bioretention and infiltration)	5. Planter boxes
		6. Biowales
		7. Permeable pavements
Water quality improvement (i.e. Reuse opportunities)	Permeable pavements	8. Constructed wetlands (Surface flow)
	Constructing wetlands	9. Constructed wetlands (surface and subsurface flow)
	Wetland restoration/ conservation	10. Treatment wetlands (subsurface flow specially)
	Constructing wetlands	11. Green walls (for greywater treatment)
Moderation of extreme events (i.e. flood)	Green roofs	1. Green roofs
	Green spaces (bioretention and infiltration)	2. Rain Gardens
		3. Planter boxes
		4. Biowales
	Water harvesting	5. Rain water harvesting systems
	Permeable pavements	6. Permeable pavements
	Constructing wetlands	7. Constructed wetlands
		12. Urban stormwater

mitigate emerging risks increases the resilience of management policies, authorities, institutions and in general the population, with the enhancement of their capacity for generating societal benefits. This requires the necessary integration of water management across sectors, policies, institutions and even geographical frameworks: talk is about an integrated and sustainable management of water resources (WWAP). The pollution generated by human activities constitutes one of the generating elements of uncertainty and risks. Data available on the effects of water pollution in Mexico, show that 4.36 deaths, 197.26 disability-adjusted life years per 100,000 population and 92,688.9 million pesos were the costs attributable to water pollution in 2004, about 0.7% of the gross domestic product (Navarro et al., 2013a, 2013b, 2013c).

One of the fundamental components of the hydrological cycle is surface water and within it the rivers. The whole of society widely benefits from the ecological services provided by rivers and lakes in a catchment area and therefore has a direct or indirect influence on them and their aquatic biota. Water use involves numerous modifications to the morphology of rivers, such as the construction of hydroelectric dams and irrigation channels. Regarding water use in basins, the main sources of point and nonpoint pollution are agriculture, industry and urbanization. The final destination of the pollutants is precisely the rivers and lakes. The effects generated by the discharge of untreated wastewater to the receiving water bodies affect these ecosystems in all its aspects and involve risks to the health of the resident population in the corresponding watersheds (Vaishali et al., 2013). Related to the aforementioned problems, water scarcity, with two components, quantity (to which greater attention has been paid), and quality is already a global problem (Junguo et al., 2016).

The risks to human, animal and ecosystem health in general, are usually related to the presence of priority pollutants and pathogens in surface water, because it is an important vehicle for transporting them. Additionally, pathogens are becoming increasingly resistant to antibiotics due to their indiscriminate use and their ubiquitous presence in the aquatic environment. Related to the latter, in the different environmental compartments and specifically in the waters of the rivers are all products that in one way or another we use in our daily lives encompassing a growing list of unregulated chemicals, known as emerging pollutants, emerging contaminants or contaminants of emerging concern, (EC). Although their concentrations in the aquatic environment are usually low, they can be considered pseudo persistent as they are continuously released into the aquatic environment. Its consequences and impacts on human health and ecosystems are not well known yet, but among them are some compounds classified as endocrine disruptors (Petrie et al., 2015; Shi et al., 2012; Du et al., 2014; Kosma et al., 2014; Neoh et al., 2016). It is a known fact that a significant group of ECs is not removed until the desired levels in the conventional treatment systems, being found in concentrations of the order of the ppb and ppt in the effluents of the WWTP (Brown, 2009b; Brown, n.d.; Buurman, 2018).

Even though all the treatment technologies are continuously being improved, conventional systems of wastewater treatment, with high consumption of energy and requirements of other inputs, constant monitoring and maintenance, are not suitable for small urban and rural communities as well as for the isolated settlers for whom it is not feasible to provide communal sanitation services. Other advanced and improving technologies still have to overcome some difficulties that make them unattractive for small communities, for example membrane fouling and energy demand for membrane reactors (Vymazal, 2010). The decentralization of sanitation and the use of green or soft water treatment technologies with a minimum requirement of energy and very low costs of operation and maintenance is the most suitable option for these sectors of the population. These low input systems which mimic natural processes of pollutant degradation and water quality restoration, have long been recognized as sources of cultural ecosystem services and other positive outcomes, not yet recognized widely enough. They make a significant contribution to sustainable development and increase benefits, including economic, to the beneficiaries of ecosystem services (Vymazal, 2013; Wu, 2015).

Treatment wetlands, also called constructed or artificial wetlands (CW), are an engineered emulation of natural wetlands and consist of a support media, plants and the circulating water. They are mainly classified according to the contact of the circulating water with the upper air layer (surface and subsurface flow) and the direction of the flow (horizontal of vertical flow) and may be the result of the combination of various types of wetlands (also with other green technologies such as treatment ponds), called hybrid systems. CW allow the removal of organic matter, nutrients, metals, EC and pathogens by a combination of biological, physical and chemical mechanisms which occur in nature. (Kadlec et al., 2009; Puigagut et al., 2007; Matamoros et al., 2012a, 2012b; Reyes-Contreras et al., 2012; Dordio et al., 2013). At the present time CW are a feasible, cost effective and environmentally friendly alternative to reducing the problems associated with water pollution. They are used to treat domestic sewage, industrial wastewater, landfill leachate, polluted river water, effluents of aquaculture systems, agricultural runoff, etc. (Puigagut, 2007) CW are suitable for small communities (below 2000 inhabitants), especially in rural areas, where the cost or energy demands of conventional and advanced treatment systems are prohibitive (Vymazal, 2015). CW have demonstrated the ability to remove the EC and other chemical pollutants as well as considerably reduce human pathogens and faecal indicators (Wu et al., 2016; Navarro et al., 2013a, 2013b, 2013c; Navarro et al., 2015; Hedegaard et al., 2014; Healy et al., 2007).

In this work, the experiences of a collective of researchers from different institutions in Colombia and Mexico in the treatment of various types of contaminated water using CW of various configurations at pilot level are addressed. The results show the effectiveness of treatment wetlands and the possibility of its use as a tool of local action to restore impaired water quality.

8.5.1 Study Sites and Pilot Systems

Izúcar de Matamoros, México. The Nexapa River Sub-basin (Fig. 8.2), NRS, is located in the Hydrologic Region 18, River Balsas (High Balsas), Atoyac basin. The most important surface water flows are the Nexapa river which rises in the eastern skirts of the volcano Popocatepetl about 20 km north of the city of Atlixco and its tributaries the Atila and Atotonilco streams. The Nexapa river is of permanent flow. fed in its upper portion by the snowmelt from the volcano. On its western portion, it receives contributions from the Cantarranas river and La Leona and Cuescomate creeks.

Through the Portezuelo channel, to the river are transferred 4 $m^3\ s^{-1}$ of heavily polluted water from the Atoyac River, after the latter receives municipal and industrial discharges from the city of Puebla. In the NRS predominates a warm humid climate, mainly in its central and southern parts, with an annual average temperature over 22 °C, with little fluctuation in the average monthly temperatures (between 5 and 7 °C). The rainy season encompasses the June to October months.

The predominant economic activity is agriculture (about 75% of the water use) and the main crops are corn, sugar cane, vegetables, gladiola and beans; another is aquaculture with tilapia and catfish farms; livestock is another economic activity

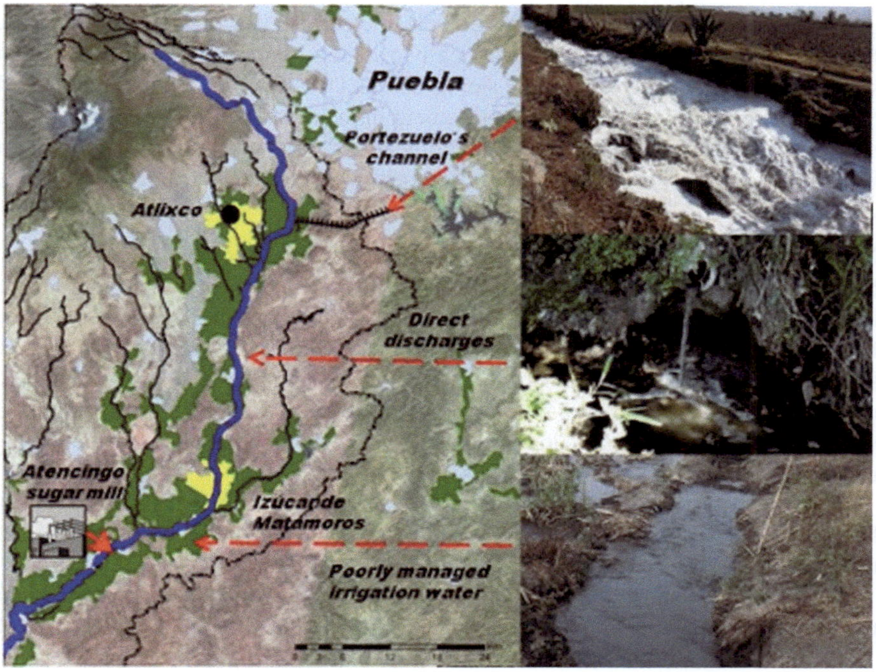

Fig. 8.2 The Nexapa River Subbasin and its more important water management problems

and, in some cases, cattle is fed a lot in housing areas; recreation and tourism with pre-Hispanic and colonial sites are part of the regional economy. The main problems concerning the pollution of waterbodies are the discharges of municipal waters to the river and the quality of the river water itself (Navarro et al., 2013a, 2013b, 2013c). So, the target of the research was the treatment of polluted river waters and the tertiary treatment of the effluent from the wastewater treatment plant of the city of Izúcar de Matamoros (WWTPIM).

Three different wetland systems using synthetic containers, whose basic features and configuration are shown in Table 8.2 and Fig. 8.3, were used for wastewater treatment: HSSF located in the WWTPIM (HSSF-M-Mx) using 15 mesocosms of 2 m^2 area, based on a modified Latin square experimental design; vertical subsurface flow wetlands (VSSF), consisting of two units of 1 m^2; 8 HSSF mesocosms of 0.6 m^2 (two unplanted as control), in the campus of the Technological University of Izúcar de Matamoros (UTIM, HSSF-R-Mx) to treat synthetic river water that emulated the average composition of the Nexapa river in terms of inorganic components, biochemical oxygen demand (BOD_5) and EC. The macrophytes *Phragmites australis* (Phr) and *Typha latifolia* (Ty) were collected near the UTIM and *Cyperus papirus* (Cyp) was acquired in greenhouses. Plants were washed and preserved and transplanted with a 10 cm distance between plants. In all cases the systems were fed with clean water until the plants reached about 40 cm and subsequently adapted to treat water for

30–40 days. Porous medium gravel and volcanic gravel (volcanic rock) were used, both available in the region. In the latter case in some systems different porous media sizes were used. All materials were sieved and its D60 was determined and washed to remove any fine material, before placement in the mesocosms. Experiments with a rustic system for the treatment of the effluent of a food packing plant and with a hybrid system (VF-VF-HSSF), are described elsewhere (Hladik et al., 2009; USEPA, 2014) (Fig. 8.3).

Monteria, Colombia. The region of Montería, Córdoba, Colombia is characterized by intense livestock activity. The cattle in the farms are bathed by spraying with a knapsack sprayer, using a pesticide. For pumping spray, lots of 7 cattle are selected for each application (20 L of water with 20 mL of the active component). Only a small portion of the mixture adheres to the skin of the animal, while the vast majority of the mixture falls to the floor of the corrals where it is mixed with faeces and urine which is periodically hosed down with water that is disposed around the farms, contaminating water bodies (Fig. 8.4).

To study the effectiveness of CW to remove the pollutants from the water of cattle baths four HSSF (HSSF-L-Co) of two stages each were installed in the campus of the University of Córdoba.

The residual water to treat was taken from a farm where the livestock is bathed using a pesticide whose active ingredient is cypermethrin. The portion that runs off

Table 8.2 Experimental setup of CW in Izúcar de Matamoros

CW & Pre treatment	Experimental Characteristics			
HSSF-M-Mx, biofilter	*Ty*-Gr-1d (D_{60}: 1.27 cm)	*Ty*-Tf-3d (D_{60}: 0.6 cm)	*Ty*-Tg-5d (D_{60}: 1.27 cm)	*Ty*-Tf-1d R (D_{60}: 0.6 cm)
	Phr-Gr-5d (D_{60}: 1.27 cm)	*Phr*-Tf-1d (D_{60}: 0.6 cm)	*Phr*-Tg-5d (D_{60}: 1.27 cm)	*Phr*-Tg-1d R (D_{60}: 1.27 cm)
	Cy-Gr-3d (D_{60}: 1.27 cm)	*Cy*-Tf-5d (D_{60}: 0.6 cm)	*Cy*-Tg-1d (D_{60}: 1.27 cm)	*Phr*-Tf-1d (D_{60}: 1.27 cm)
VF-M-Mx, biofilter	Depth (D_{60}): 30 cm (coarse sand), 30–45 cm (0.2–0.8 cm), 45–55 cm (0.5 a 2.0 cm), 55–70 cm (2.0 a 4.0 cm), 70–80 cm, (8.0 -10.0 cm) Plants: Ty and Phr. 4 pulses of 40 L in 8h, system resting 16h			
HSSF-R-Mx, none	HRT 1, 3 y 5d			
	***Ty*-Gr** (D_{60}: 1.0 cm)	***Ty*-Tg** (D_{60}: 0.4 cm)	***Phr*-Gr** (D_{60}: 1.0 cm)	***Phr*-Tg** (D_{60}: 0.4 cm)
	Cy-Gr (D_{60}: 1.0 cm)	**Cy-Tg** (D_{60}: 0.4 cm)	**Gr** (D_{60}: 1.0 cm)	**Tg** (D_{60}: 0.4 cm)

Note HRT–hydraulic residence time; d–days; Gr–Gravel; T–tezontle, coarse (g) and fine (f)

Fig. 8.3 CWs experimental setup at Izúcar de Matamoros: **A** HSSF in the WWTPIM **B** inlet and **C** outlet details; **D** VF in the WWTPIM; **E** HSSF in UTIM

Fig. 8.4 Cattle bath problem in Montería, Colombia. **A** Spraying with pesticides; **B** Corral washing; **C** Polluted pond near a corral. *Source* (Map downloaded from http://www.cordoba.gov.co/v1/cordoba_mapa_politico.html)

from the floor during the bathing and corral hosing was collected. This initial sample was stored in plastic tanks and moved immediately to the location of the CW pilot system. As pre-treatment (Fig. 8.3), a sieve was used to retain the solids of greater size and a system of filters of sand and gravel with descending flow in the first filter and upward in the second filter, in order to reduce the number of suspended solids. The effectiveness of sand filters in the removal of pesticides has been demonstrated and in combination with CW for the treatment of dairy wastewaters (Ferguson, 2013b; Fletcher, 2015). Subsequently the effluent of the filters was led to a system of two 1000 L plastic tanks interconnected and with recirculation to maintain the greatest possible consistency throughout the experiment.

The filtered residue was supplied through PVC piping to 8 mesocosms of 80 × 80 × 40 cm (long-high-width) arranged in four rows with a pair of modules each. 3 rows were planted with the different macrophytes and the fourth, without plants, served as a control (Fig. 8.3). The modules were filled with gravel of three different sizes, placed, top-down from smaller to larger sizes. On each unit of two modules a flow of water was regulated for a hydraulic residence time of 5 days. The flow was revised 2 twice a day to regulate its homogeneity in the 4 units. In each module a device for sampling was placed at the output.

The studied species, papyrus (*Cyperus papyrus*), yellow lily (*Limnocharis flava*) and ginger *Alpinia purpurata* (sp), were collected in the department of Cordoba, on sites where they usually grow, selecting only those plants that could support a process of transplantation and without an excessive size (not more than 14 cm). All plants were adapted for a month, in containers with gravel similar in size to the mesocosms, selected and transplanted subsequently to the mesocosms those that showed better development (an average height of 30 cm, 10 plants per module). Before starting the removal experiments, the system was adapted and stabilized for 30 days with a flow of the residual water similar to that used in further experiments.

Medellín, Colombia. One problem common to many localities in Latin America, is the indiscriminate dumping of solid waste. The work done in the Morro de Moravia, allowed the experimental evaluation of the effectiveness of CW to treat the leachate of those wastes. The Morro de Moravia was a landfill located within the city limits of Medellín. Having been occupied for several years by an irregular settlement, it was rescued and now presents a completely different look (Fig. 8.6). However, under the restored areas are buried tons of garbage of all kinds that cause hazardous leachates. To solve this problem a system of vertical flow wetlands was constructed (Fig. 8.7).

The treatment system is composed as follows:

- 110 m of PVC perforated line of Ø3" and Ø2" as horizontal filter.
- Input housing in fiberglass diameter 0.6 m and depth 0.6 m.
- Sedimentation tank and up flow anaerobic filter in fiberglass with a 1600 L capacity.
- Fiberglass distribution box of 2 m in length, 0.6 m wide and 0.6 m high, with six compartments.
- Valve system for operation and maintenance.
- 12 VF wetlands (VFVF-LL-Co) of fiberglass 1.20 m in diameter and 1.50 m high.

Fig. 8.5 Pre-treatment system and pilot CWs at University of Córdoba

Fig. 8.6 The Morro de Moravia (Medellin, CO). Before and after restoration (A and B respectively)

Fig. 8.7 CW in the Morro de Moravia, Medellín. **A** Drainage system; **B** Sedimentation tank; **C** VF units

- Collection box of treated water in fiberglass of 2.00 m in diameter and 1.50 m high, with cover.
- Cylindrical polyethylene infiltration box of 500 L with a cover of alfajor foil.

The leachate from the Morro, after a first purification step in buffer strips placed in its skirts was treated in the described system of CW (Fig. 8.7).

8.5.2 Analytical Methods and Statistical Analysis

To assess the efficiency of operation of the treatment systems, samples of the influent and effluent of the different experimental units were periodically taken and the physical–chemical determined based on Mexican and Colombian standards or their equivalent APHA-AWWA or USEPA methods.

For organic compound analysis the water samples were extracted and analysed by GC/MS following the EPA SW-846, EPA 8270d in Izúcar de Matamoros () and the USGS method 5-C2 for the analysis of cypermethrin in Montería (Reif et al., 2011). Extraction and clean-up of the samples followed the recommendations on the mentioned methods. Prior to extraction techniques, when necessary, all samples were spiked with surrogates. For analysis the samples were reconstituted with an internal standard. Previous to injection in Izúcar de Matamoros the samples were derivatized with trimethylsulfonium hydroxide. Recoveries and concentrations of the target compounds were evaluated with six-point calibration curves of authentic standards ($R^2 > 0.95$).

ANOVA was used to evaluate the experimental results (Statistica 12, StatSoft, Tulsa, OK, USA, in Izúcar de Matamoros and StatGraphics Centurion 16 for

Windows, Statpoint Technologies, Inc. Warrenton, Virginia, USA, in Montería). Differences were considered significant when $p < 0.05$.

8.5.3 Results and Discussion

Izúcar de Matamoros. To demonstrate the effectiveness of the CW in Izucar the Matamoros, in this work the BOD5 was selected as representative of the total organic load of the polluted waters and the fragrance Galaxolide (GAL) as an example of EC. For the analysis were considered the loads of the two pollutants, expressed as units of mass per unit area and day. In the case of the rustic system only BOD5 was analysed.

As can be seen in Table 8.2, the input loads in the case of the WWTP, as was to be expected, were superior to those of the experiments conducted in UTIM, where was simulated the composition of the water of the river, with an average BOD_5 of 100 mgL^{-1}. The loads of the industrial waters were highly variable, a characteristic aspect of these processes.

With regard to the studied variables, the HRT has a significant influence on all systems ($p < 0.05$, Table 8.3 and Fig. 8.8), for the removal of BOD_5 and GAL. There was no significant influence of the macrophyte nor of the porous medium in the HSSF-M-Mx. In the UTIM, the inclusion of unplanted systems, contributes to a significant influence of the macrophyte. This allows for the conclusion that the macrophytes play an important role in the removal of the studied pollutants.

In the HSSF-M-Mx, the results observed in BOD_5 removal for HRT of 1, 3 and 5 days were respectively (indicating the minimum and maximum values and the macrophyte with better results): 76.9–89.5, *Cyp*; 71.5–90.8, *Ty*; 64.9–76.5, *Cyp*. For the removal of GAL: 80.8–96.5%, *Cyp*; 82.0–93.2%, *Ty*; 62.8–76.6%, *Phr*. In a similar manner in the HSSF-R-Mx for BOD_5 are observed the following results: 61.6–84.8%, *Cyp*; 34.5–77.3%, *Ty*; 30.0–57.0%, *Phr*. For Gal: 85.6–96.2%; 81.2–94.7%; 48.7–81.1%, being the results with *Ty* the best obtained. In all cases, the unplanted

Table 8.3 Input loads to the CW systems

System	HRT	Galaxolide, g m^{-2} d^{-1}	BOD_5, g m^{-2} d^{-1}
		Input load	Input load
HSSF-M-Mx	5	179–373	12.6–26.3
	3	164–250	11.6–17.7
	1	506–664	35.7–46.8
VF-M-Mx	na	221.3 ± 84.9	40.8 ± 6.8
HSSF-R-Mx	5	76–119	2.6–4.0
	3	161–178	3.8–5.5
	1	150–183	9.2–11.2
HSSF-FP-Mx	10	Na	1194 ± 207 to 3650 ± 990

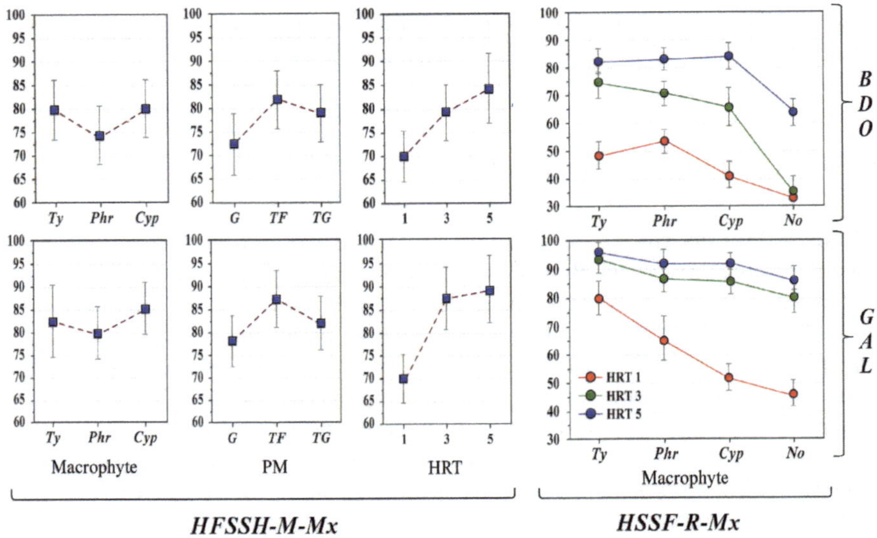

Fig. 8.8 Effect of the studied variables in the removal percentages in Izúcar de Matamoros. Vertical bars indicate the 95% confidence intervals

systems showed the smallest removals. The results obtained in these HFSSH CW show that the three macrophytes contribute to the removal of the studied pollutants, being the *Ty* the most efficient with the smaller HRT. The results obtained in the removal of GAL are comparable with those reported in other studies (* MERGEFORMAT Table 8.4). In the case of the VF-M-Mx, were observed %removals of BOD_5 and GAL of 43.7, 46.1; 71.0, 81.4, with *Phr* and *Ty* respectively. These results are comparable with those reported by (Ogata Y et al., 2015) in 2011 (* MERGEFORMAT Table 8.4) and show the feasibility of its use for the tertiary polishing of the WWTPIM effluent.

The removal of the BOD_5 in the treatment of the water from vegetables packing process, with a very high organic load, reported previously (USEPA, 2014) 94 and 40% for loads of up to 1200 and around 2200 g m^{-2} d^{-1}, indicate the importance of establishing an anaerobic pre-treatment before the CW to equalize the variable BOD_5 loads observed for industrial waters. This load and flow equalization can be

Table 8.4 Main effects ANOVA of the HSSF CW results in Izúcar de Matamoros

	HFSSH-M-Mx		HFSSH-R-Mx	
	BOD_5	Galaxolide	BOD_5	Galaxolide
Macrophyte	0.213394	0.495179	***0.000002***	***0.000395***
PM	0.107361	0.154554	0.794320	0.955731
HRT	***0.019125***	***0.005149***	***0.000000***	***0.000000***

accomplished in the case of domestic waters with a septic tank upstream of the CW (Diem Trang et al., 2010).

Concerning the hybrid systems, the overall removal of COD and Gal were 87–91 and 95–99% without significant differences between the two combinations of macrophytes used (*Phr-Phr-Ty* & *Phr-Ty-Ty*). COD removal is comparable to those obtained in French two stages VF CW, 86 ± 7% (Molle et al., 2015). The turbidity diminished from 70–105 to 2.7–4.3 NTU so clear and odourless water (indols and coprostanol were efficiently removed, data not shown) was obtained at the end of the hybrid system. Total coliforms decreased by 2–3 orders of magnitude, *E. Coli* levels by 1–2 orders of magnitude, Salmonella by 2–3 orders of magnitude and Vibrio cholera was completely removed. These results illustrate how water quality is improved using CW so less "aggressive water" can be discharged to the surface water bodies.

The observed absence of significant differences between treatments with different porous media and macrophytes in the studied CW shows that they can be combined giving flexibility in the use of the available local materials for the CW in the study region.

Montería Colombia. The first remarkable thing (Table 8.5) is that the pre-treatment decreases between 37 and 50% the high values of the COD, total phosphorus (TP), total suspended solids and the concentration of cypermethrin (Ccyp) of the cattle bathing waters. This may be related to the association of the substances that contribute to these values with the particulate phase of the raw residual water.

In the mesocosms, good removals of the target pollutants were achieved (* MERGEFORMAT Table 8.6). TP removals were above 80% for all cases, even for the unplanted bed, with average outflow concentrations between 2 and 4 mg L^{-1}, for an average inlet concentration 19.37 ± 2.04 mg L^{-1}. These results indicate a high retention of phosphorus within the system and its association with the particulate phase (Table 8.7).

Table 8.5 GAL of removal in CW

Removal, %	Process	HRT d	Macrophyte	References
46–89	Biological filtration	4		Galvão and Matos (2012)
30–50	HSSF	2.3	*TY*	García and Corzo (2008)
85–95	HSSF	8.5	*Ty and Phr*	Dolman et al. (2013)
~ 67	Restored CW	From 3 to 20	*Phr,* *Ty and Schoenoplectus*	Ghermandi and Fichtman (2015)
76–88	HFSSH	2.5	*Phr*	Dordio and Carvalho (2013)
11–65	VF	na	*Phr*	Gkika et al. (2015)

Table 8.6 Characterisation of the raw corral filtration water

Water	pH	COD (mg L^{-1})	TP (mg L^{-1})	TSS (mg L^{-1})	C_{cyp} (mg L^{-1})
Untreated	7,45	1996.8	31.05	141.46	13.25
Pre-treated	7,15	998.4	19.37	75.1	8.25
% Removal		50.0	37.6	46.9	37.7

Table 8.7 Initial values and removal percentages of the studied variables

Inflow value (mg L^{-1})	TP 19.37 ± 2.04	TSS 74.38 ± 6.87	COD 1004.7 ± 86.3	Ccyp 8.25 ± 0.21
Macrophyte	Removal, %			
Papyrus	83.4 ± 0.02a,b	62.4 ± 8.7a	59.4 ± 5.8a	90.7 ± 7.9a,b
Yellow lily	85.0 ± 0.06a	50.1 ± 18.4a,b	69.1 ± 3.7b	97.9 ± 2.5a
Gínger	85.1 ± 0.02a	57.5 ± 13.6a,b	62.9 ± 5.0a,b	89.3 ± 8.4a,b
Unplanted	81.0 ± 0.03b	37.1 ± 7.7b	63.5 ± 4.5a,b	83.9 ± 7.5b

Note Different letters indicate significative differences ($p < 0.05$)

The removal of TSS was between 37.1 and 62.4%, with lower results in the unplanted bed. The fact that *Cyperus papyrus* has the higher TSS removal may be associated with development of its root zone, since this plant presents large cylindrical and horizontal rhizomes, which can grow in different directions, rather than with the plant metabolism. Thus, a simple mechanical action can generate the hydraulic conditions that favour greater efficiency in the elimination of TSS in the Papyrus mesocosms.

The highest removal of COD was found for the wetland planted with yellow lily, while the lowest were observed in the Papyrus mesocosm. In descending order of removal: Yellow Lily > Unplanted > Gínger > Papyrus. The mean outflows Ccyp are between 0.7 and 1.4 mg L^{-1} of cypermethrin, for an inlet concentration of 8.25 mg L^{-1}. The observed average removals vary from 83 to 98 with the highest values found in the mesocosm planted with yellow lily. In descending order of average removal: Yellow Lily > Papyrus > Gínger > Unplanted.

The obtained results are very important for the studied region as very high concentrations of this pesticide have been found, 31.05 ± 0,41 and 13.42 ± 0,42 $\mu g\ L^{-1}$ in summer and winter respectively, in the irrigation channels of the agricultural district of la Doctrina [unpublished, 38], one of the final destinations of the surface waters of the watershed that receives the cattle bathing water.

Medellín, Colombia. After the systems start-up, by causes not related to its operation, the flow of leachate ceased, so here are presented the results of a single analysis of the system's performance. Nevertheless, it reflects the ability of the system to remove some of the pollutants present in the influent (Table 8.8). The VF-LL-Co removed with effectiveness the ammoniacal nitrogen and the biodegradable organic load, with a low percentage of removal for COD. An effluent of better colour and lower turbidity

Table 8.8 Physical and chemical parameters of the influent and effluent from the VF-LL-Co

Parameter	COD mg L^{-1}	BOD5 mg L^{-1}	N-NH3 mg L^{-1}	Turbidity, NTU	Colour, CVU
Inflow	236.6	148	237.44	5.97	341.538
Effluent	193.1	18	43.54	1.23	253.077
% Removal	18.4	87.8	81.7	Na	Na

was obtained, with results comparable to those reported in the literature (Garrone et al., 2016). This demonstrates the potential application of these systems for the decentralized treatment of the leachate from accumulations of solid waste that are found indiscriminately near small communities in the Latin American context.

8.6 Conclusions

The resilience of small rural communities in the face of the situation imposed by climate change and increasing migration towards large urban centres is an essential step to strengthen its population through a more dignified life. Sustainable water management is a key component of this, it is necessary to consider not only the technological aspects of the technology but also those which are set out below.

It is an incontrovertible fact that the CW have been successfully used for decades for the treatment of domestic and other types of wastewater in temperate regions in greater proportion and in an increasing proportion in tropical countries, where better results should be obtained due to the warmer climate, as seen, for example, in HSSF in Viet Nam and in the vertical French wetlands used with success in French-speaking countries (World Bank, 2006; Méndez-Fajardo et al., 2011). Their advantages with regards to the costs are still demonstrated (Wang et al., 2016) So, why is the technology not sufficiently widespread in Latin American countries and remains as necessary its demonstration as a first stage of any community intervention? The answer is not simple and, in its essence, covers the complex economic, social, and political processes that are manifested in each community and that in the long run threaten the local capacity for the sustainable management of water and sanitation, regardless of how strong the government regulation is in this regard. The interests of the groups of power, the electoral circumstances, land-owner interests-conflicts and the ignorance of the consequences of the inadequate management of water in the current climate situation as a background, these facts fragment the community in face of the challenge of water (Padilla-Rivera et al., 2016) If it is wanted to ensure that the community feels and takes possession of the technology for the treatment of their waste water, an aspect that the experience demonstrates, it is necessary that this need permeates the population and that any external actor, considers with humility its social, ethnic, environmental and geographical context, leading to various forms of conceiving of water treatment (Dalahmeh et al., 2009; Muga et al., 2008).

The problem does not end in the social aspects. There is another difficulty to solve. The constant improvement of the treatment technologies, both existing and those that are in the stages of development, means that there are multiple treatment options when the evaluation of alternatives for a project of this type is initiated. It is necessary to consider all the options of treatment because each community may require a different solution and other soft technologies, such as the stabilization/oxidation ponds, which are competitive or reasonably combinable with the CW (Kalbar et al., 2012) This selection process is complex, because it is necessary to consider not only the technical-environmental and economic components but is increasingly important to consider the social component in any community intervention (Molinos-Senante et al., 2014).

Indisputably, is not only a problem from the multidisciplinary technical and economic points of view but also a multisector one from the social aspects, from design to deployment, where each sector (academy, authorities, communities), have a specific contribution to the materialization of the technology (Molinos-Senante et al., 2015) This explains the growing interest in the development of indicators and methodologies that allow the selection of the most appropriate and sustainable technology for different scenarios (Gkika et al., 2015; Kalbar et al., 2016) Various aspects which can be considered are weighed in each component by pair or multiple comparison of the alternatives (with a good rating for the CW) and already these algorithms have been embodied in a software (Cao et al., 2016; de la Varga et al., 2015; López et al., 2015).

Additionally, it is necessary to further explore the use of local materials, macrophytes, porous medium and waterproofing materials, which contribute to reducing the ecological footprint of the CW (Aguilar et al., 2015) Similarly, in this same sense is necessary to study the emission of greenhouse gases depending on aspects of design and operation of these systems.

Although not studied in the present work, it is necessary to mention, as a contribution of great value to mitigate non-point pollution to surface water, the usefulness of the CW for the treatment of the runoff from the agricultural fields, an aspect that is a problem in the context of the rural and even urban sites of the studied regions.

The CW studied in this work have shown to be efficient in the tertiary polishing of the effluent from a wastewater treatment plant, surface water of similar composition to that of a heavily polluted river, residual water of corrals for cattle management and leachates from urban solid waste. These types of wastewaters are frequent in the field of small rural communities in Latin America, so CW constitute a tool for local action in sustainable water management in the mentioned human settlements.

For the implementation of this and other soft technologies that contribute to the resilience of small rural communities, it is necessary to follow a bottom-up schema, with the population of the community as promoter and manager of the idea and its realization.

It is necessary to continue with the studies about the assessment of the most appropriate technologies for each locality and assess the ecological footprint of the systems so as not to create one problem when solving another.

Acknowledgements The authors would like to thank the entities that financed the projects in the framework of which the studies were conducted and that may be consulted in the references and sites of the participating entities. Also are grateful to the Institutions of Higher Education and the municipal authorities involved for the material and moral support to the investigations.

References

Aguilar, L., Gallegos, A., Arias, C.A., Caro, P., Sahuquillo, S., Pérez, C., & Morató, J. (2015). Regeneration of irrigation water and reuse of runoff and drainage water in agricultural plots by constructed wetlands. In *Wetpol 2015—6th International Symposium on Wetland Dynamics and Control Annual Conference of the Constructed Wetland Association.* 13th to 18th September, 2015, York, UK.

Arias, M., & Brown, M. (2009). Feasibility of using constructed treatment wetlands for municipal wastewater treatment in the Bogot_a Savannah, Colombia. *Ecological Engineering 35*, 1070e1078. https://doi.org/10.1016/j.ecoleng.2009.03.017

Ashley, R., Lundy, L., Ward, S., Shaffer, P., Walker, L., Morgan, C., Saul, A., Wong, T., & Moore, S. (2013). Water-sensitive urban design: Opportunities for the UK. In *Proceedings of the Institution of Civil Engineers: Municipal Engineer*, vol. 166, no. 2, pp. 65–76. https://doi.org/10.1680/muen.12.00046

Barron, N. J., Kuller, M., Yasmin, T., Castonguay, A. C., Copa, V., Duncan-Horner, E., Gimelli, F. M., Jamali, B., Nielsen, J. S., Ng, K., Novalia, W., Shen, P. F., Conn, R. J., Brown, R. R., & Deletic, A. (2017). Towards water sensitive cities in Asia: An interdisciplin-ary journey. *Water Science and Technology, 76*(5), 1150–1157. https://doi.org/10.2166/wst.2017.287

Bernhardt, E., Zandaryaa, S., Arduino, G., Jimenez-Cisneros, B., Payne, J., Zadeh, S., McClain, M., Irvine, K., Acreman, M., Cudennec, C., Amerasinghe, P., Dickens, C., Cohen-Shacham, E., Fedotova, T., Cox, C., Bertule, M., Coates, D., Connor, R., Simmons, E., Gastelumendi, J., & Gutierrez, T. (2018). NBS [Nature-based solutions] for managing water quality. In *WWAP (United Nations World Water Assessment Programme); UN-water*. The United Nations World Water Development Report 2018: Nature Based Solutions for Water (Paris, France).

Brown, R. R., Keath, N., & Wong, T. H. F. (2009b). Urban water management in cities: Historical, current and future regimes. *Water Science and Technology, 59*(5), 847–855. https://doi.org/10.2166/wst.2009.029

Brown, R. R., Keath, N., & Wong, T. H. F. (2009a). *Urban water management in cities: Historical, current and future regimes* (pp. 847–855). https://doi.org/10.2166/wst.2009.029

Brown, R., Rogers, B. & Werbeloff, L. (n.d.). *Moving toward water sensitive cities and policy makers.*

Buurman, J., & Padawangi, R. (2018). Bringing people closer to water: Integrating water management and urban infrastructure. *Journal of Environmental Planning and Management, 61*(14), 2531–2548. https://doi.org/10.1080/09640568.2017.1404972

Cao, W., Wang, Y., Sun, L., Jiang, J., & Zhang, Y. (2016). Removal of nitrogenous compounds from polluted river water by floating constructed wetlands using rice straw and ceramsite as substrates under low temperature conditions. *Ecological Engineering, 88*(2016), 77–81.

Council of Australian Governments—COAG (2004). *Council of Australian Governments' meeting [1] 25 June 2004: COAG communiqué.* Canberra: Council of Australian Governments.

Dalahmeh, S. S., Assayed, M., & Suleiman W. T. (2009). Themes of stakeholder participation in greywater management in rural communities in Jordan. *Desalination, 243*, 159–169.

de la Varga, D., Ruiz, I., Álvarez, J. A., & Soto, M. (2015). Methane and carbon dioxide emissions from constructed wetlands receiving anaerobically pretreated sewage. *Science of the Total Environment, 538*, 824–833.

Deeptha, V. T., Sudarsan, J. S., Baskar, G. (2015). Performance and cost evaluation of constructed wetland for domestic waste water treatment. *Journal of Environmental Biology, 36*(5), 1071–1074.

Dobbie, M. F., Brown, R. R., & Farrelly, M. A. (2016). Risk governance in the water sensitive city: Practitioner perspectives on ownership, management and trust. *Environmental Science and Policy, 55*, 218–227. https://doi.org/10.1016/j.envsci.2015.10.008

Dobre, C. C. (2013). Urban storm management: Possible cat-alyst for moving towards a water sensitive city. *Archi DOCT, 1*(July), 42–56.

Dolman, N., Savage, A., & Ogunyoye, F. (2013). Water-sensitive urban design: Learning from experience. *Proceedings of the Institution of Civil Engineers: Municipal Engineer, 166*(2), 86–97. https://doi.org/10.1680/muen.12.00033

Dordio, A. V., & Carvalho A. J. (2013) Organic xenobiotics removal in constructed wetlands, with emphasis on the importance of the support matrix. *Journal of Hazardous Material, 252–253*, 272–292.

Du, B., Price, A. E., Scott, W. C., Kristofco, L. A., Ramirez, A. J., Chambliss, C. K., Yelderman, J. C., & Brooks, B. W. (2014). Comparison of contaminants of emerging concern removal, discharge, and water quality hazards among centralized and on-site wastewater treatment system effluents receiving common wastewater influent. *Science of the Total Environment, 466–467*, 976–984.

Everard, M., & McInnes, R. (2013). Systemic solutions for multi-benefit water and environmental management. *Science of the Total Environment, 461–462*, 170–179.

Faivre, N., Fritz, M., Freitas, T., de Boissezon, B., & Vandewoestijne, S. (2017). Nature-Based Solutions in the EU: Innovating with nature to address social, economic and environmental challenges. *Environmental Research, 159*, 509e518. https://doi.org/10.1016/j.envres.2017.08.032

Farrelly, M., & Brown, R. (2011). Rethinking urban water management: Experimentation as a way forward? *Global Environmental Change, 21*(2), 721–732. https://doi.org/10.1016/j.gloenv-cha.2011.01.007

Ferguson, B. C., Brown, R. R., Frantzeskaki, N., de Haan, F. J., & Deletic, A. (2013a). The enabling institutional context for integrated water management: Lessons from Melbourne. *Water Research, 47*(20), 7300–7314. https://doi.org/10.1016/j.watres.2013.09.045

Ferguson, B. C., Frantzeskaki, N., & Brown, R. R. (2013b). A strategic program for transitioning to a water sensitive city. *Land-Scape and Urban Planning, 117*, 32–45. https://doi.org/10.1016/j.landurbplan.2013.04.016

Fletcher, T. D., Shuster, W., Hunt, W. F., Ashley, R., Butler, D., Arthur, S., Trowsdale, S., Barraud, S., Semadeni-Davies, A., Bertrand-Krajewski, J. L., Mikkelsen, P. S., Rivard, G., Uhl, M., Dagenais, D. & Viklander, M. (2015). SUDS, LID, BMPs, WSUD and more—The evolution and application of terminology surrounding urban drainage. *Urban Water Journal, 12*(7), 525–542. https://doi.org/10.1080/1573062X.2014.916314

Floyd, J., Iaquinto, B. L., Ison, R., & Collins, K. (2014). Managing complexity in Australian urban water governance: Transitioning Sydney to a water sensitive city. *Futures, 61*, 1–12. https://doi.org/10.1016/j.futures.2014.04.002

Folke, C. (2006). Resilience: The emergence of a perspective for social-ecological systems analyses. *Global Environmental Change, 16*(3), 253–267. https://doi.org/10.1016/j.gloenv-cha.2006.04.002

Galvão, A., & Matos, J. (2012). Response of horizontal sub-surface flow constructed wetlands to sudden organic load changes. *Ecological Engineering, 49*, 123–129.

García, J., & Corzo, A. (2008). Purification with constructed wetlands. Practical Guide for design, construction and exploitation of subsurface flow wetland systems. Departamento de Ingeniería Hidráulica, Marítima y Ambiental. Universidad Politécnica de Catalunya, España, pp. 2–10.

Garrone, P., Grilli, L., Groppi, A., & Marzano, R. (2016). Barriers and drivers in the adoption of advanced wastewater treatment technologies: a comparative analysis of Italian utilities. *Journal of Cleaner Production, 135*, 1300–1311.

Ghermandi, A., & Fichtman, E. (2015). Cultural ecosystem services of multifunctional constructed treatment wetlands and waste stabilization ponds: Time to enter the main stream? *Ecological Engineering, 84*, 615–623.

Gkika, D., Gikas, G. D., & Tsihrintzis V. A. (2015). Environmental footprint of constructed wetlands treating wastewater. *Journal of Environmental Science and Health Part A Toxic/Hazardous Substances and Environmental Engineering, 50*(6), 631–638.

De Haan, F. J., Rogers, B. C., Frantzeskaki, N. & Brown, R. R. (2015). Transitions through a lens of urban water. *Environmental Innovation and Societal Transitions, 15*, 1–10. https://doi.org/10.1016/j.eist.2014.11.005

He, B.-J., et al. (2019). Co-benefits approach: Opportunities for implementing sponge city and urban heat island mitigation. *Land Use Policy, 86*, 147–157.

Healy, M. G., Rodgers, M., & Mulqueen, J. (2007). Treatment of dairy wastewater using constructed wetlands and intermittent sand filters. *Bioresource Technology, 98*(12), 2268–2281.

Hedegaard, M. J., & Albrechtsen, H. J. (2014). Microbial pesticide removal in rapid sand filters for drinking water treatment–potential and kinetics. *Water Research, 48*, 71–81.

Hladik, M. L., Smalling, K. L., & Kuivila, K. M. (2009). *Methods of analysis—Determination of pyrethroid insecticides in water and sediment using gas chromatography/mass spectrometry: U.S. geological survey techniques and methods 5–C2*, 18 p.

Hoyer, J., Dickhaut, W., Kronawitter, L., & Weber, B. (2011). Water sensitive—urban design: Principles and inspiration for sustainable stormwater management in the city of the future; [elabo-rated in the Context of the Research Project SWITCH—Managing Water for the City of the Future].

Jenkins, M. (2020). *Sponge city: Shenzhen explores the benefits of designing with nature*. Lincoln Institute of Land Policy. April 2.

Junguounguo, L., Qingying, L., & Hong, Y. (2016). Assessing water scarcity by simultaneously considering environmental flow requirements, water quantity, and water quality. *Ecological Indicators, 60*, 434–441.

Kadlec, R. H., & Wallace, S. D. (2009). *Treatment Wetlands* (2nd ed., p. 1016). CRC Press.

Kalbar, P. P., Karmakar, S., & Asolekar, S. R. (2012). Selection of an appropriate wastewater treatment technology: A scenario-based multiple-attribute decision-making approach. *Journal of Environmental Management, 113*, 158–169.

Kalbar, P. P., Karmakar, S., & Asolekar, S. R. (2016). Life cycle-based decision support tool for selection of wastewater treatment alternatives. *Journal of Cleaner Production, 117*, 64–72.

Kosma, C. I., Lambropoulou, D. A., & Albanis, T. A. (2014). Investigation of PPCPs in wastewater treatment plants in Greece: Occurrence, removal and environmental risk assessment. *Science of the Total Environment, 466–467*, 421–438.

Leigh, N. G., & Lee, H. (2019). Sustainable and resilient urban water systems: The role of decentralization and planning. *Sustainability (Switzerland), 11*(3). https://doi.org/10.3390/su11030918

Lloyd, S., Wong, T., & Blunt, S. (2012). Water-sensitive cities: Applying the framework to Melbourne. *Australian Journal of Water Resources, 16*(1), 65–74. https://doi.org/10.7158/W10-834.2012.16.1

López, D., Fuenzalida, D., Vera, I., Rojas, K., & Vidal, G. (2015). Relationship between the removal of organic matter and the production of methane in subsurface flow constructed wetlands designed for wastewater treatment. *Ecological Engineering, 83*, 296–304.

Lundqvist, J., Turton, A., & Narain, S. (2001). Social, institutional and regulatory issues. In C. Maksimovic & A. Tejada-Guibert (Eds.), *Frontiers in urban water management: Deadlock or hope?* (pp. 344–398). IWA Publishing.

Marrugo, J., Arteaga, P., & Sánchez, C. (2013). Pesticides in irrigation channels of the agricultural distric of la Doctrina (Córdoba-Colombia). Unpublished results.

Mas, I. F., Rizzo, A., & Regelsberger, M. (2018). The role of constructed wetlands in a new circular economy, resource oriented, and ecosystem services paradigm. *Journal of Environmental Management, 216*, 275–284.

Masi, F., Rizzo, A., & Regelsberger, M. (2018). The role of constructed wetlands in a new circular economy, resource oriented, and ecosystem services paradigm. *Journal of Environmental Management, 216*, 275–284.

Matamoros, V., Sala, L., & Salvadó V. (2012). Evaluation of a biologically-based filtration water plant. *Bioresource Technology, 104*, 243–249.

Matamoros, V., Arias, C. A., Nguyen, L. X., Salvadó, V., & Brix H. (2012). Occurrence and behavior of emerging contaminants in surface water and a restored wetland. *Chemosphere, 82*, 1083–1089.

Matamoros, V. Y., & Salvadó, V. (2012). Evaluation of the seasonal performance of a water reclamation pond-constructed wetland system for removing emerging contaminants. *Chemosphere, 86*, 111–117.

Méndez-Fajardo, S., Opazo, M., Rosemary, Y., & Perez, B. C. (2011). Methodology for the appropriation of technologies of basic sanitation in indigenous communities. *Notebooks of Rural Development, 8*(66), 153–176.

Molinos-Senante, M., Gómez, T., Garrido-Baserba, M., Caballero R., & Sala-Garrido R. (2014). Assessing the sustainability of small wastewater treatment systems: A composite indicator approach. *Science of the Total Environment 497–498*, 607–617.

Molinos-Senante, M., Gómez, T., Caballero R., Hernández-Sancho, F., & Sala-Garrido, R. (2015). Assessment of wastewater treatment alternatives for small communities: An analytic network process approach. *Science of the Total Environment, 532*, 676–687.

Molle, P., Latune, R. L., Riegel, C., Lacombe, G., Esser, D., & Mangeot, L. (2015). French vertical-flow constructed wetland design: Adaptations for tropical hot climates. *Water Science and Technology 71*(10), 1516–1523.

Muga, H. E., & Mihelcic, J. R. (2008). Sustainability of wastewater treatment technologies. *Journal of Environmental Management, 88*, 437–447.

Navarro, A., Herrera, J., Caso, L., & Marrugo J. (2013). Water quality of the river Nexapa: Spatio-temporal trends and their implications. In M. Ramos., V. Aguilera. (eds.), *Natural and exact science* (pp. 83–97). Handbook-©ECORFAN-Valle de Santiago, Guanajuato.

Navarro, A., Philp, R., Marrugo, J., & Herrera J. (2013). Emerging contaminants in different environmental compartments. In: González-Pérez, J. A., González-Vila, F. J., Jiménez-Morillo, N. T., & Almendros G. (Eds.), *Organic geochemistry, trends for the 21th century, vol. 1. Book of abstracts of the communications presented to the 26th international meeting on organic geochemistry* (pp. 372–373). Spain: Costa Adeje, Tenerife.

Navarro, A., Morales, L., Sánchez, J., & Montoya, J. (2015). Hybrid wetlands for removal of organic micropollutants from municipal wastewaters. *Revista Latinoamericana el Ambiente y las Ciencias, 6*(12), 86–90.

Navarro, A., García, Y., Vázquez, A., & Marrugo, J. (2013a). Efficiency of a subsurface flow wetland for the treatment of wastewater from the vegetables packing. *Avances En Ciencias e Ingeniería, 4*(1), 39–50.

Neoh, C. H., Noor, Z. Z., Ahmad Mutamim, N. S., & Lim, C. K. (2016). Green technology in wastewater treatment technologies: Integration of membrane bioreactor with various wastewater treatment systems. *Chemical Engineering Journal, 283*, 582–594.

Noyola, A., Padilla-Rivera, A., Morgan-Sagastume, J.M., Güereca, L.P., & Hernández-Padilla, F. (2012). Typology of municipal wastewater treatment technologies in Latin America. *Clean – Soil, Air, Water, 40*(9), 926–932.

O'Hogain, S., & McCarton, L. (2018). *A Technology portfolio of nature based solutions: Innovations in water management*. Springer.

Ogata, Y., Ishigaki, T., Ebie, Y., Sutthasil, N., Chiemchaisri, C., & Yamada, M. (2015). Water reduction by constructed wetlands treating waste landfill leachate in a tropical region. *Waste Management, 44*, 164–171.

Özerol, G., Dolman, N., Bormann, H., Bressers, H., Lulofs, K., & Böge, M. (2020). Urban water management and climate change adaptation: A self-assessment study by seven midsize cities in the North Sea Region. *Sustainable Cities and Society, 55*(September 2019), 102066. https://doi.org/10.1016/j.scs.2020.102066

Padilla-Rivera, A., Morgan-Sagastume, J., Güereca, L., & Hernandez-Padilla, F. (2012). Typology of municipal wastewater treatment technologies in Latin America. *Clean 40*, 926e932. https://doi.org/10.1002/clen.201100707

Padilla-Rivera, A., Morgan-Sagastume, J. M., Noyola, A., & Güereca, L. P. (2016). Addressing social aspects associated with wastewater treatment facilities. *Environmental Impact Assessment Review, 57*, 101–113.

Paing, J., Guilbert, A., Gagnon, V., & Chazarenc, F. (2015). Effect of climate, wastewater composition, loading rates, system age and design on performances of French vertical flow constructed wetlands: A survey based on 169 full scale systems. *Ecological Engineering, 80*, 46–52.

Petrie, B., Barden, R., & Kasprzyk-Hordern B. (2015). A review on emerging contaminants in wastewaters and the environment: Current knowledge, understudied areas and recommendations for future monitoring. *Water Research 72*, 3–27.

Puigagut, J., Villaseñor, J., Salas, J. J., Bécares, E., & García, J. (2007). Subsurface-flow constructed wetlands in Spain for the sanitation of small communities: A comparative study. *Ecological Engineering, 30*, 312.

Reif, C., Besancon, A., le Corre, K., Jefferson, B., Lema, J., & Omil, F. (2011). Comparison of PPCPs removal on a parallel-operated MBR and AS system and evaluation of effluent post-treatment on vertical flow reed beds. *Water Science and Technology, 63*(10), 2411–2417.

Reyes-Contreras, C., Hijosa-Valsero, M., Sidrach-Cardona, R., Bayona, J. M., & Bécares, E. (2012). Temporal evolution in PPCP removal from urban wastewater by constructed wetlands of different configuration: A medium-term study. *Chemosphere, 88*(2), 161–167.

Reyes-Contreras, C., Matamoros, V., Ruiz, I., Soto, M., & Bayona, J. M. (2011). Evaluation of PPCPs removal in a combined anaerobic digester-constructed wetland pilot plant treating urban wastewater. *Chemosphere, 84*(9), 1200–1207.

Rijke, J., Farrelly, M., Brown, R., & Zevenbergen, C. (2012). Configuring transformative governance to enhance resilient urban water systems. *Environmental Science and Policy, 25*(ii), 62–72. https://doi.org/10.1016/j.envsci.2012.09.012

Rijke, J., Farrelly, M., Brown, R., & Zevenbergen, C. (2013). Configuring transformative governance to enhance resilient urban water systems. *Environmental Science and Policy, 25*(ii), 62–72. https://doi.org/10.1016/j.envsci.2012.09.012

Rogers, B. C., Dunn, G., Hammer, K., Novalia, W., de Haan, F. J., Brown, L., Brown, R. R., Lloyd, S., Urich, C., Wong, T. H. F., & Chesterfield, C. (2020). Water sensitive cities index: A diagnostic tool to assess water sensitivity and guide management actions. *Water Research, 186*, 116411. https://doi.org/10.1016/j.watres.2020.116411

Shi, H., Cheng, X., Wu, Q., Mu, R., & Ma, Y. (2012). Assessment and removal of emerging water contaminants. *Journal of Environmental & Analytical Toxicology S2*, 003. https://doi.org/10.4172/2161-0525.S2-003

Diem Trang, N. T., Konnerup D., Schierup, H., Chiem, N. H., Tuanc, L. A., & Brix, H. (2010). Kinetics of pollutant removal from domestic wastewater in a tropical landscape subsurface flow constructed wetland system: Effects of hydraulic loading rate. *Ecological Engineering, 36*, 527–535.

USEPA. (2014). Method 8270d. Semivolatile organic compounds by gas chromatography/mass spectrometry, Revision V.

Vaishali, P., & Punita, P. (2013). Assessment of seasonal variation in water quality of River Mini, at Sindhrot, Vadodara. *International Journal of Environmental Science, 3*(5), 1424–1436.

Vymazal, J. (2010). Constructed wetlands for wastewater treatment. *Water, 2*, 530–549.

Vymazal, J. (2013). The use of hybrid constructed wetlands for wastewater treatment with special attention to nitrogen removal: A review of a recent development. *Water Research 47*(14), 4795–4811.

Vymazal, J., Březinová, T., & Koželuh, M. (2015). Occurrence and removal of estrogens, progesterone and testosterone in three constructed wetlands treating municipal sewage in the Czech Republic. *Science of the Total Environment 536*,625–631.

Wang, X., Tian, Y., Zhao, X., Wu, Q., Peng, S., & Yan, L. (2016). Optimizing the operation of combined oxidation pond-constructed wetland ecosystems used for treating composite wastewater. *Ecological Engineering, 88*, 64–76.

Wells, M. J. M., Pellegrin, M. L., Morse, A., Bell, K. Y., & Fono L. J. (2009). Emerging Pollutants. *Water Environmental Research 81*(10), 2211–2254.

Wong, T. H. F., & Brown, R. R. (2009a). *The water sensitive city: Principles for practice* (pp. 673–682). https://doi.org/10.2166/wst.2009.436

Wong, T. H. F., & Brown, R. R. (2009b). The water sensitive city: Principles for practice. *Water Science and Technology, 60*(3), 673–682. https://doi.org/10.2166/wst.2009.436

World Bank. (2006). Biofilter: A sustainable option for the treatment of waste water in small localities. Washington, DC: World Bank. https://documents.worldbank.org/curated/en/2006/04/6771338/biofiltro-una-opcion-sostenible-para-el-tratamiento-de-aguas-residuales-en-pequenas-localidades

Wu, H., Zhang, J., Ngo, H. H., Guo, W., Hu, Z., Liang, S., Fan, J., & Liu, H. (2015). A review on the sustainability of constructed wetlands for wastewater treatment: Design and operation. *Bioresource Technology, 175C*, 594–601.

Wu, S., Carvalho, P. N., Müller, J. A., Manoj, V. R., & Dong, R. (2016). Sanitation in constructed wetlands: A review on the removal of human pathogens and fecal indicators. *Science of the Total Environment, 541*, 8–22.

WWAP (World Water Assessment Program). (2012). *The United Nations world water development report 4: Managing water under uncertainty and risk.* Paris: UNESCO.

Chapter 9
Social Technologies for City Resilience: Lessons from Vulnerable Areas in Brazil

Andréa Cardoso Ventura

Abstract The discussions about the necessity to make cities more resilient to climate events point out these spaces as "centres of technology and innovation". In fact, much knowledge is being developed in cities all around the world. The aim of this paper is to highlight the role of simple and intensive knowledge technologies developed in Brazil, using the accumulated experience from vulnerable communities in its semi-arid region. They have developed and implemented extensive strategies that have enabled them to reduce their vulnerability to past climate variability and change, in association with social movements. The experiences have taken place under a social movement called the "Convivência com o Semiárido" (coexistence with the semi-arid), with the use of Social Technologies (ST), considering traditional or high technological solutions (occidental traditional knowledge) inappropriate for the social, environmental and economic context of the semi-arid. The objective of this paper is to present some of these experiences, revealing how these communities are using accumulated traditional knowledge to become more resilient to climate change. The study was based on multiple study cases and in-depth interviews and concludes that ST can have a key role to play in a close and yet uncertain future.

Keywords Traditional knowledge · Social movement · Semi-arid regions · Climate change

9.1 Introduction

Especially in the last 15 years, much has been discussed about the necessity to make cities more resilient to climate events. These preoccupations result, among other actions, in the creation by the United Nations Office for Disaster Risk Reduction (UNISDR) of the "Making Cities Resilient" Campaign, in 2010. The campaign was created to examine the very important role of cities as "centres of technology and innovation" for more than a half of the world population. Its objective is raising

A. C. Ventura (✉)
Federal University of Bahia, Salvador, Brazil
e-mail: andreaventurassa@gmail.com

J. Morato et al. (eds.), *Intangible Cultural Heritage and Traditional Knowledge Related to Urban Resilience*, Resilient Cities, https://doi.org/10.1007/978-3-032-12713-6_9

awareness and commitment for sustainable development practices that will reduce disaster risk and increase the wellbeing and safety of citizens" (UNISDR, 2010, p. 3).

One of the milestones of this action was the establishment of a "Ten Essentials for Making Cities Resilient checklist", to help city leaders and local governments to work alongside local activists, grassroots networks and national authorities. The UNISDR consider that this guide can assist local governments and other agencies to implement the "Hyogo Framework for Action 2005–2015: Building the Resilience of Nations and Communities to Disasters", adopted by 168 governments in 2005.

What has happened, as highlighted by Nyong et al. (2007), local populations around the world through their traditional knowledge systems, have developed and implemented extensive strategies that have enabled them to reduce their vulnerability to past climate variability and change, strategies which sometimes have exceeded those predicted by models of future climate change. If local knowledge is to have an impact beyond local or national levels, it is necessary to accumulate evidence from local studies in order to be relevant to global negotiations and permeate the negotiations (Burton et al., 2007). Furthermore, a great range of circumstances and different strategies offer an opportunity for learning more about causes, consequences and alternative strategies at a local scale (Wilbanks, 2003), that can help in city resilience creation.

This is certainly the case of cities located in semi-arid regions in Brazil, characterized by severe and frequent droughts. People living in these regions have had a long history of living and surviving climate variations. Based on traditional knowledge and in association with social movements, they have developed diverse strategies to reduce their vulnerability and increase their resilience to environmental change over the years. There have been a large number of experiences trying to reverse, or at least to minimize the vulnerability of Brazilian semi-arid regions.

As highlighted by Carvalho (2010), most of these experiences are being developed under a social movement called the "Convivência com o Semiárido" (coexistence with the semi-arid), that has been trying to find practical local solutions to face semi-arid climate change vulnerability since the 1980's. This movement is an umbrella term for different networks, local ideologies and attitudes with the common aim of helping vulnerable communities to adapt to live under semi-arid climate conditions. Most of the strategies proposed and developed within this movement involve the use of Social Technologies (ST), given that traditional or high technological solutions (occidental traditional knowledge) are not appropriate for the social, environment and economic context of the semi-arid (Santos, 2011). This accumulated knowledge is being considered a possible key to making these places more resilient to future risks that are being expected in the region as a consequence of climate changes (Fernandez et al., 2016; Ventura et al., 2014).

With the aim of contributing to research on resilience to climate change on a local scale, this paper explores some of the ways in field research and in-depth interviews with relevant stakeholders, look at strategies based on local knowledge developed to reduce vulnerability to climate change. The aim of this paper is to present the experience of cities in Brazilian semi-arid regions to develop what is called ST.

It is expected that, in the end of this article, the reader can see that the ST are helping these communities in, at least, the following 4 of the 10 essential actions to be taken to make their cities more resilient as determined by UNISDR (2010, 2012):

- Check List Point 1. Put in place organisation and coordination to understand and reduce disaster risk, based on participation of citizen groups and civil society. Build local alliances. Ensure that all departments understand their role in disaster risk reduction and preparedness.
- Check List Point 4. Invest in and maintain critical infrastructure that reduces risk, such as flood drainage, adjusted where needed to cope with climate change.
- Check List Point 6. Apply and enforce realistic, risk compliant building regulations and land use planning principles. Identify safe land for low-income citizens and upgrade informal settlements, wherever feasible.
- Check List Point 8. Protect ecosystems and natural buffers to mitigate floods, storm surges and other hazards to which your city may be vulnerable. Adapt to climate change by building on good risk reduction practices.

The paper, based on doctoral research and some publications deriving from the study, begins by reviewing the semiarid environment in Bahia. The next section shows the concept of ST and describes why ST projects are considered potential instruments to link poverty and the fight against climate change, promoting the resiliency of the cities in Brazilian semi-arid regions. Section 9.4 presents the research method. Section 9.5 introduces and discusses the results; and Sect. 9.6 summarizes the major arguments in the paper.

9.2 The Semi-arid Region in Brazil: Droughts and Perspectives with Climate Reality

Brazil is well-known around the world as home to one of the richest forests in the world: the Amazon Forest. The abundance of water and biodiversity is one of the most common associations made with the country. However, what many people are unaware of is that Brazil has one of the most highly populated dry areas on the planet: its semi-arid region. The Brazilian semi-arid region covers an area of 1,797,940 km^2 and is located in the Northeast of the country. It covers 8 north-eastern states and part of the state of Minas Gerais. This vast region is inhabited by 28% of Brazil's population (around 20 million people) (Brazil, 2011).

The IPCC (2007) has identified the semi-arid regions around the world as one of the most vulnerable regions to climate impacts. In the Brazilian semi-arid region, desertification is currently threatening about 15% of the territory, a reduction in rain is expected as a consequence of global warming; and average temperatures could rise by 4 °C in a pessimistic scenario, and from 2 to 3 °C in the most optimistic scenario (Santos, 2008).

Nobre et al (2004), in a wide study on the climatic consequences of large-scale desertification in north eastern Brazil, identified that desertification is a consequence of a reduction in rainfall, and an increase in evapotranspiration. Unfortunately, desertification and inappropriate land use are threatening about 15% of the region (Leal et al., 2005).

More recent studies confirm the projections, suggesting an increase in dryness in the region, with rainfall reductions, temperature increases and water deficits and longer dry spells (Brito et al, 2018; Marengo & Bernasconi, 2015). In fact, the region faced a drought from 2012 to 2015 with an intensity and impact not seen in decades (Marengo et al, 2017).

The Brazilian semi-arid region presents the worst Human Development Index in the country, associated with an absence of social infra-structure (such as access to water and basic sanitation) (Carvalho, 2010). The region is historically known for the socio-economic setbacks and agricultural failures caused by dry spells and severe droughts associated with deficits in political-institutional capacities (Lindoso et al., 2011). Moreover, as the impacts of climate change, drought and desertification are closely interlinked, populations whose livelihoods depend principally on natural resources acutely suffer from them (Stringer et al, 2009).

Despite this context, people living in the Brazilian semi-arid region have had a long history of living and surviving with climate variations. Based on traditional knowledge and in association with social movements, they have developed diverse strategies to reduce their vulnerability and increase their resilience to environmental change over the years. There have been a large number of experiences trying to reverse, or at least to minimize the vulnerability of the Brazilian semi-arid region. As said before, the social movement called the "Convivência com o Semiárido" (coexistence with the semi-arid) has been trying to find practical local solutions to face semi-arid climate change vulnerability since the 1980's, involving the use of ST.

The present research was conducted in the state of Bahia, considered by Toni and Holanda (2008) to be one of the driest areas in the Brazilian Northeast region. Bahia is the 5th largest Brazilian state occupying 36.33% of the Northeast and its semi-arid region represents 69.34% of its total area (Bahia, 2010). Like other semi-arid regions in Brazil, long term drought and desertification, combined with poor socio-economic indicators have had adverse social, economic and environmental consequences in the Bahia semi-arid region, such as water scarcity, agricultural loss, threatened biodiversity, disease and forcing urban migration (Barbieri et al, 2010; Stringer et al, 2009). Nonetheless, Bahia is an economically significant region for Brazil in terms of its agro-economic potential (Bahia, 2011; Santos, 2008). This is partly because communities have assessed, adapted, experienced and invented new ways to deal with the drought and the scarcity of water (Barbosa, 2010).

9.3 Social Technology: The Role of Knowledge and Participation in the Path of Development

Technologies are social constructs. Their implementation plays a central role in the processes of social change, determining the positions and behaviours of actors and social distribution structures, production costs, and access to goods and services (Ilha & Ribeiro, 2012).

Since the 1970s, there have been many attempts to develop "non-conventional" technologies. The most representative movement claiming the development of a technology with a "human face" was the Intermediate Technology Development Group (ITDG), founded by the economist E.F Schumacher, author of the widely influential book "Small Is Beautiful" (Schumacher, 1973). Schumacher developed his vision of what he called an "intermediate" technology, known now as Appropriate Technology (AT), which emphasizes technology as people-centred, focusing on the use of sustainable technologies as a poverty solution and a way to promote development in developing countries. Thus, the concept of AT was based on the dissemination of technologies in developing countries that do not imply large capital investment nor advanced technology while allowing for industrial and agricultural development. However, according to Dagnino (1976), in most cases, AT was introduced from developed countries with the associated problem that development models of advanced countries could not work in countries with lower levels of economic development. This weakness was tolerated because for a long time, technology was considered to be culturally "neutral", but nowadays it is widely recognized that it is crucial for the transfer processes to pay particular attention to the technological assimilation of communities and to the incorporation of such technology into their cultural and social habits (Narváez, 1996).

As the AT movement did not admit to this necessity of empowering local actors and including their knowledge and culture within technology design, a wave of critical movements grew up in the 1980s.

The concept of Social Technology (ST) ("Tecnologias Sociais", in Portuguese), was developed in Brazil during the decade of 2000 by a network of researchers, social movements and organizations, as a response to the criticisms of AT. The most important difference that distinguishes ST from AT is that, in ST the process of generation and diffusion of alternative knowledge occurs through the participation of local actors (Dagnino et al., 2004). Although there is no official definition for the concept of ST, Thomas (2009) defined it as a way to design, develop, implement and manage technology oriented to solve social and environmental problems, generating social and economic dynamics of social inclusion and sustainable development. ST concept recognizes popular wisdom and scientific knowledge, acting as a bridge between social demands and solutions through the application of local knowledge (ITS, 2007). Commonly, ST projects are understood as products, techniques or methods that can be replicated developed and/or applied in interaction with a community, which represents social transformation solutions through the sustainable use of local resources (RTS, 2010). For ST, replication implies that it must be recreated in new places, appropriated to their reality, bringing new knowledge and meanings (Miranda, 2011).

The replication process is crucial within the ST concept as is part of the scaling up (Gillespie, 2004). Scaling up small-scale success projects is said to lead to more quality benefits for more people over a wider geographical area, more quickly, more equitably and more enduringly (Franzel et al, 2004). However, it has been criticized that scaling up often only scales up the formula and sometimes the process is forgotten (Commonwealth Secretariat, 2005). ST projects present a different paradigm, involving many small-scale projects at the local level instead of one large, remote project, and so involve a lot of stakeholders. Thus, ST concept recognizes that ST projects may succeed in generating sustainable development of vulnerable communities only because they operate in micro contexts. Furthermore, the focus of the replication process is not the replication of specific elements of a given technology, but the process that allowed the project to succeed and above all the local roots and conditions that can sustain the change.

The movement towards ST development and adoption is increasing yearly in Brazil (RTS, 2010). Nevertheless, research exploring the role that ST could take to face climate change is new. Some recent literature has warned about the potential of ST to face climate change (Santos, 2011; Ventura et al., 2011, 2012; Banco do Brazil Foundation, 2012), but new empirical accounts are required to explore the expected synergies between adaptation and mitigation while promoting human development (Ventura et al., 2012). Three main issues support the selection of ST as a mechanism to link adaptation and mitigation while having the potential to be influential in the design of the post-2012 regime. First, the wide range of actors involved. NGOs, universities, government agencies, and social movements exchange knowledge within a participatory process in the ST project cycle (Ashby, 2003). The inclusion of different actors in the decision-making process has recently been highlighted as key to effectively addressing climate change, either for adaptation or mitigation or adaptation and mitigation (Laukkonen et al., 2009). Second, there is the possibility of ST replication to different places with similar social and environmental problems (Dagnino et al., 2004; RTS, 2010). As previously stated, the possibility of up scaling is key in order to be relevant to global negotiations (Burton et al., 2007). And third, the wide range of techniques and resources covered under different areas such as assistive technologies, local development of communities, and food safety. Likewise, ST covers different socio-economic and natural resources such as energy, water availability and human health.

9.4 Methodology

This paper identifies local strategies for dealing with climate change impacts in the Bahia semi-arid region and analyses their potential to face climate change. In order to carry out the analysis, a database for the ST projects in the Bahia semi-arid area (as of September 2011) was created. To identify the ST projects, a database developed by the Núcleo de Política e Administração em Ciência e Tecnologia of the Bahia Federal University (NACIT, 2010) was primarily used. Information in NACIT´s database was completed with ST projects that were accepted with some of the ST grants recognized in Brazil. Therefore, the data use ST collected until 2013.

The updated database used for this analysis covers 130 projects. Firstly, it was necessary to identify which of the ST projects were developed in the Bahia semi-arid region. The locations of the projects were cross-referenced with the list of municipalities in the Bahia semi-zone, as proposed by Ministério da Integração Nacional (BRASIL, 2005). 60 of the 130 projects were developed in the Bahia semi-arid area. Then, the projects, which incorporated mitigation, adaptation or both options, were identified, given that, adopting these strategies ST is contributing to resilience creation.

To identify mitigation or adaptation strategies, and to select the natural and socio-economic resources impacted by such strategies, the options identified by Fernandez et al. (2012) served as a basis to construct the framework of analysis of this paper (Table 9.1).

The documents of the 60 ST projects were reviewed in detail looking for potential impacts on confronting climate change. Then, 10 projects were selected by contact

Table 9.1 Potential strategies option to face climate change

Resource	Strategies options
Water	• Water storage techniques • Develop renewable-based water pumping systems • Basic sanitation • Take climate factors into consideration to manage resources
Agriculture	• Adjustment of planting dates and crop variety • Crop rotations • Soil conservation or recuperation • Enhance energy efficiency irrigation system for agriculture • Post-harvest food processing (for food security) • Management of agroforestry systems for enhancing resource use efficiency and crop productivity • Reduction of agricultural fires • Composting of organic waste with agricultural purposes • Dedicated energy crops to replace fossil fuel use
Energy	• Reducing dependence on fossil fuel • Renewable heat and power • Improve energy efficiency
Residue	• Biogas utilization from animal defecation • Recycling or reutilization practices • Waste treatment
Forestry	• Reduced deforestation • Protection to natural forestry and biodiversity • Agroforestry management to improve resource use efficiency and productivity

Source adapted from Fernandez et al. (2012)

facility and convenience to receive visits onsite and extensive interviews with relevant stakeholders were conducted.

9.5 Results and Discussion

The analyses carried out with the 60 ST identified in the Bahia semi-arid region found that 27 projects contributed with direct impact on climate change through mitigation or adaptation options. These technologies presented at least one of the potential strategies presented in. A resume of these technologies can be found on. All the ST signalled with an asterisk were visited onsite.

As is possible to notice from observing the table above, most of the ST are related to water and agriculture. It is something to expect from the ST, since they are developed to provide solutions in a critical environment and to deal with social problems in the region. The scarcity of water generates important risks for food production. Though, the community and its organisations mobilize themselves to create possible solutions to the problem, using resources that are locally available, like the cisterns used in the Water and Citizenship Program in the Bahia semi-arid area.

It is extremely important to point out that, even more than contributing to making the localities more resilient to the future climate conditions, the analysed ST are contributing to improving many aspects of local life conditions.

All the 10 experiences studied in loco presented governance between local actors, efficient use of natural resources and promotion of economic sustainability. Most of them generated new ways of creating income for the population. Studies carried out by this researcher (Ventura et al, 2014) concluded that these ST are really promoting endogenous technology development and transfer, promotion of training to users, at the same time that they are enabling the promotion of food and nutritional security.

Unfortunately, this same study concludes that few of the experiences are creating capacities to face future climate events, since they are not using this knowledge to promote discussions on critical aspects of desertification and climate risks. According to Polhmann (2011), the empowerment of communities, done by its strengthening of decision-making, is a critical factor in guaranteeing the effectiveness of climate change resilience at the local level. Field studies suggest that a wide range of stakeholder participation strategies is used during the ST project cycle. Such strategies include different degrees of participation ranging from consultation to empowerment through collaboration and co-decision, as defined by Luyet et al (2012). Participation is identified by the Commonwealth Secretariat (2005) as one of the most promising practices for scaling up successful projects. It is necessary to move forward on this subject.

For most ST projects, participation techniques combine presentations and public hearings, interviews, field visits and interactions, workshops and consensus conferences. This way of involving local stakeholders has achieved mainstreaming of local knowledge in public policies, and therefore scaling up.

The best example of this is the use of cisterns for rainwater collecting through family initiatives. Nowadays the technology is a public policy called "Programa Um Milhão de Cisternas" (One Million Cisterns Program). Interviewed actors highlighted that the adequate management of the programs is a challenge to all the stakeholders involved and that popular participation in the local development process is pivotal to achieving the win–win potential between the climate and development interfaces. This ST is a good example of successful replication, together with other technologies, like the subterranean barrage (Photo 3, below), board walk cistern, popular water pump and polycultures (Photo 4), that shows empirical scaling up potential. In fact, local visits and interviews demonstrated that all the visited projects were small-scale projects being replicated with the involvement of a wide range of stakeholders from other successful small-scale projects in Brazilian semi-arid regions (Figs. 9.1 and 9.2).

Besides this, the integration of traditional and scientific knowledge must be improved. Only 5 of the 10 visited projects involved universities or research centres to upgrade the technology. And in only 3 of them was the local government involved.

Fig. 9.1 Photos 1 and 2: Rainwater collection system, using cisterns

Fig. 9.2 Photo 3: Duffle installation; Photo 4: Polyculture beside an area with subterranean barrage in a semi-arid area. *Source* Photo 3: ASA (2011); *Source* Photo 4: Own archive

Considering the importance of reapplication of these solutions, it would be very interesting if the ST developers would associate themselves with other institutions to improve and up-scale the results.

Despite these points, the visited ST projects proved a serious capacity to help solve environmental and social issues related to climate change. As can be seen in Table 9.2, most of the ST projects include more than one product, technique or methodology to provide solutions in the local spectrum, allowing self-management and execution of work (i.e., combination of crop relocation, with energy efficiency irrigation systems and soil conservation strategies).

Another important verification must be highlighted. Links between agriculture, development and climate change are complex. According to the FAO (2009), agriculture directly releases large quantities of greenhouse gases into the atmosphere amounting to around 14% of global annual anthropogenic GHG emissions.

If indirect contributions (e.g., fertilizer production and distribution or land conversion) are factored in, the contribution of agriculture could be as high as 17–32%. According to CGEE (2009), 75% of Brazilian CO_2 emissions came from changes in land use, especially from the conversion of forests into grazing land.

However, the results suggest that the examined ST projects for agriculture provide both mitigation and adaptation benefits. This is because the traditional local model of agricultural production in the semi-arid areas based on principles that create healthy soils and cultivate biological diversity while dismissing the use of chemical fertilizers and enhancing the use of energy efficient strategies such as manual water pumps for irrigation systems. A good example of how an agriculture experience can be good for both sides is found in ST "Agro ecological Integrated and Sustainable Production" (Photo 5). The technology is a circular garden where vegetables and leaves are grown and kept by the local community (Fig. 9.3).

This project consists of farming with simple techniques and integration already known by many rural communities, who embrace a culture that relies on the least possible external farming inputs, promoting mainly food sustainability in small rural properties. The main techniques used are: drip irrigation, poultry, small animals, goats, cattle, according to local / regional suitability. The droppings of these animals are the raw material for production of compound fertilizer, which is used in plantations. Production is quite diverse to take full advantage of soil nutrients, control pests and diseases. This adds more value to family income through the production of fruits, roots and pasture for small animals. This type of production system allows the rational use of water, the protection of riverbanks reducing the surface runoff and maintaining the soil rich in organic matter and mulch, which facilitates infiltration of rainwater into soil. This ST saves diesel while reducing the transportation of agricultural inputs, since most of the supplies come from the property itself and products are consumed by its producing families.

Table 9.2 Social technologies with potential to contribute to face climate change

Sector(s)	ST project	Description
Agriculture	Social alliance	Utilization of a social educative and participatory methodology to get community involvement in social and environmental problem solutions. Emphasis on agro ecology techniques
Agriculture and water	Subterranean barrage*	Implantation of a subterranean system to collect and store rainwater, using a duffle installation. The water is used for food production and for animals
	Popular water pump*	Manual water pump installation, with a gyratory lever, to reach the water in deep wells
	Rhea pump	Manual water pump installation (tubular system), to reach the water in deep wells
	Rain water collection to irrigation	Technology knowledge transfers to allow water storage for human consumption and low-cost irrigation systems
	Boardwalk cistern*	Installation of a boardwalk area to collect water. The system is linked to a cistern for storage and the water is used for food production and animals
	Scholar cisterns	Installation of two combined receivers to collect and store water. One for human consumption and another for food production and contextualized education
	São Francisco River integrated development	Training in hydrological resource management and renewable energy, associated with agro ecological production
	Solar "Pintadas"—adaptation to climate change*	Research through learning-by-doing process for new irrigation techniques for hard water. Among these, the use of solar pumps and special drop irrigation system. The use of specific techniques for food production
	Polycultures in semi-arid*	Development of agricultural practices adapted to the semi-arid region. Use of differentiated crop techniques side-by-side, appropriate irrigation systems, and food storage
	Agro ecological integrated and sustainable production*	Food production in a fully integrated agro ecological system using drop irrigation systems
	Convivial—survival in semi-arid program	Training for farmer families in the use of adaptive techniques of food production (water issues, reforesting, burning reduction and water storage systems)
	Family farming agroforestry systems*	Development of agricultural practices appropriate for the semi-arid zone, mixing arboreal plants and crops. Dissemination of techniques of cultivation and irrigation and food storage

(continued)

Table 9.2 (continued)

Sector(s)	ST project	Description
Agriculture	Food and nutritional safety in São Francisco region	Development of agro ecological techniques, diffusion of appropriate and safe food information. Training in agro ecological production
	"Sertão Vivo"—simple sanitation system	Development of a simplified sanitation system
Energy and agriculture	Biodiesel production: society, education and technology integration	Inclusion of family agriculture in the production of castor oil for biodiesel production, normally dominated by large scale farms
	Solar dryer	Development of simplified solar drying system to preserve food
Energy and waste	Collective bio digestor*	Goats' defecation utilization to biogas production. The gas is used in substitution of fossil fuel. Waste is used as organic fertilizer
Energy	Ecological cook stoves*	Substitution of rudimentary stoves for more efficient ones. The new stoves use less wood and are smoke free
Forestry	Licuri productive chain	Role of actions to collect and use the licuri (a fruit) in a sustainable manner. Evolves machine production to extract seed and to process food
	Exploration of Umbu in food preparation	Development of techniques in the use of umbu (a very perishable fruit) to juice and jelly fabrication
	Sete Passagens State park	Formalization of a state park for water origin protection and for forestry corridor creation
	"Sempre Viva" project	Planned actions to preserve the "Sempre Viva" (Syngonanthus mucugensis giulietti), native plant from semi-arid areas. Techniques to cultivate the plant (until then, only collected) to guarantee the species preservation
Waste	Integrated and sustainable development	Sanitarian actions for a treatment lake and land field installation
Waste and forestry	Agroindustry waste transformation	Transformation of the liquid waste from sisal (plant) fibber defibrillation in industrial products
Water	Water and citizenship program in Bahia semi-arid*	Rooftop rainwater harvesting system for human consumption. The technology uses a manual water pump to pump the collected water replacing electric ones
Water and waste	Sanitation, environmental education and citizenship in Pintadas	Creation of articulation spaces for strategy formulation and development of integrated strategies, such as sanitation systems, waste management, water storage and environmental education

Source own elaboration based in document analysis

Fig. 9.3 Photo 5—production unit. *Source* Semiárido show (2011)

9.6 Conclusions

Those cases of ST presented show that there is a lot of examples of local knowledge that can be used to make cities more resilient to climate change. This paper presents different local strategies in the Bahia semi-arid region, and offers an opportunity for learning more about causes, consequences and alternative strategies at a local scale.

The qualitative analysis of the ST projects suggests that the success of the experiences and its successful reapplication as a possible resilience strategy depend on both technological and management aspects. Findings suggest that it is necessary to approach the scaling-up process qualitatively, focusing on processes and not just replicating outcomes. The strategy used by the ST projects, based on the replication of small-scale projects at the local level involving a wide range of stakeholders, appears to be a successful approach that could be further explored and considered for upscaling other local initiatives.

Returning to the points presented by UNISDR (2010) as essential actions for making their cities resilient, which this researcher believes makes a strong possible contribution to the ST, it was possible to observe:

- Check List Point 1. The ST put in place good organisation and coordination to understand and reduce a climate problem (the drought) that represents a disaster risk. But it needs an improvement in participation of citizen groups and civil society to better understand the future risks and the difference between existing water and food scarcity, and the possible situation in the future. Based on concrete and consolidated knowledge, these communities could build effective local alliances that would contribute to the understanding of each role in disaster risk reduction and preparedness.

- Check List Point 4. In the Bahia semi-arid region there is a real investment in critical infrastructure that reduces risk, especially considering the risk of extreme water scarcity. The ST that are promoting the construction of cisterns in houses and schools for human consumption, and those that are working on water storage for food production and animal treatment can play an important role in this sense. But, if the local governments and the local population are not aware of the gravity of the risks, they will not be able to create a quantity and quality of these infrastructures which is sufficient to face what may happen in the near future. The actual structures are not enough even to face droughts like the one from 2012 to 2015, when cities suffered a real impact. What can be said about events that will very likely exacerbate this situation?
- Check List Point 6. The application of realistic land use planning principles must be urgently scaled up to prevent these cities from food scarcity and guarantee food safety. The ST studied proves its capacity to produce much more food than in a traditional land use form, and it is usually associated with the observation and selection of the most appropriate crop, and with the independence of external resources. But this knowledge is still restricted to those producers that were reached by ST projects. The necessity for a huge scale-up in quantity of this kind of production, respecting, of course, the process that will guarantee respect for the ST, is urgent. Together with water and agriculture technologies, those linked with energy efficiency and waste recycling or reuse, could play a very important role in upgrading low-income communities and upgrading informal settlements.
- Check List Point 8. Most of the ST projects studied are promoting the protection of ecosystems and natural buffers. They can have a crucial role in mitigating floods (in moments when there is an intensification and concentration of rain in very short periods), in a soil not prepared for it. This protection of the semi-arid ecosystem also can have a fundamental positive impact on soil conservation and recuperation to act as a food safety promotion strategy, with good risk reduction practices.

Studies on how to solve ST fragilities and promote its scaling process will be very welcome to city resilience in vulnerable areas such as semi-arid regions.

As also pointed out by a Brazilian governmental publication (BRASIL, 2016), Brazil is one of the countries with the biggest number of municipalities enrolled on the UNISDR campaign for city resilience. But simple enrolment is not enough to guarantee its resilience. Local governments must assume commitment to it, through actions and priorities. Observation and making use of existing local knowledge can make all the difference in the results.

References

ASA—Articulação do Semi-Árido Brasileiro. (2011). O Lugar da Convivência na Erradicação da Extrema Pobreza: Reflexões e proposições da Articulação no Semiárido Brasileiro (ASA), no intuito de contribuir para a garantia plena do acesso à água para todas as pessoas no Semiárido. Recife/PE. Available at http://www.asabrasil.org.br/Includes/VersaoImpressao.asp?COD_CLIPPING=1125. Access in 20 set. 2012.

Ashby, J. (2003). Introduction: Uniting science and participation in the process of innovation—Research for development. In: Pound, B., Snapp, S., Mcdougall, C., & Braun, A. (Ed.), *Managing natural resources for sustainable livehoods: Uniting science and participation*. London: Earthscan Publications Ltd. and the Internacional Development Research Center.

Bahia (2010). A Bahia no Nordeste e no Brasil: Indicadores Econômicos Comparados 2002–2009, Vol. 3. Salvador: SEI—Superintendência de Estudos Econômicos.

Bahia. (2011). Lei n. 12.050, de 07 de janeiro de 2011. Institui a Política sobre Mudança do Clima do Estado da Bahia. Published in Diário Oficial de 07 de janeiro de 2011. Available at http://www.saeb.ba.gov.br/vs-arquivos/HtmlEditor/file/sustentabilidade/LeiEstn%C2%BA12050de07dejaneirode2011.pdf. Access in 30 mai. 2011.

Banco do Brasil Foundation. (2012). Água e Mudanças Climáticas: Tecnologias Sociais e Ação Comunitária. Silva, Milton Nogueira da (Comp.). Belo Horizonte: Cedefes e Fundação Banco do Brasi. Available at http://www.sinquisp.org.br/pdfs/Livro%20Agua%20e%20Mudancas%20Climaticas%206fev12.pdf. Access in 14 Feb. 2013.

Barbieri, A. F., Domingues, E., Queiroz, B. L., Ruiz, R. M., Rigotti, J. I., Carvalho, J. A. M., & de Resende, M. F. (2010). Climate change and population migration in Brazil's Northeast: scenarios for 2025–2050. *Population and Environment, 31*(5), 344–370.

Barbosa, I. A. V. (2010). A Governamentalidade e o Desenvolvimento Internacional: Um Estudo de Caso do Acordo do Nordeste de 1962. (Dissertação de Mestrado). Programa de Pós-Graduação em Relações Internacionais do Instituto de Relações Internacionais da PUC-Rio, (92 p.). Rio de Janeiro.

Brasil. (2005). Ministério da Integração Nacional. Relatório Final Grupo de Trabalho Interministerial para Redelimitação do Semi-Árido Nordestino e do Polígono das Secas. Brasília.

Brasil. (2011). Ministério da Ciência e Tecnologia (MCT). In: Lima, R. da C. C., Cavalcante, A. M. B., & Marin, A. M. P. (Ed.), Desertificação e Mudanças Climáticas no Semiárido Brasileiro (209 p.). Campina Grande, Instituto Nacional do Semiárido—INSA.

Brasil. (2016). Ministério da Integração Nacional. Construindo cidades resilientes. Published in April 03, 2016. Modified in June 21, 2016. Available at http://www.integracao.gov.br/web/guest/cidades-resilientes?inheritRedirect=true. Access in April 01, 2018

Brito, S. S. B., Cunha, A. P. M. A., Cunningham, C. C., Alalá, R. C., Marengo, J. A., & Carvalho, M. A. (2018). Frequency, duration and severity of drought in the semi-arid Northeast Brazil region. *International Journal of Climatology, 38*(2), 517–529.

Burton, I., Bizikova, L., Dickinson, T., & Howard, Y. (2007). Integrating adaptation into policy: Upscaling evidence from local to global. *Climate Policy, 7*(4), 371–376.

Carvalho, L. D. (2010). Ressignificação e Reapropriação Social da Natureza: práticas e programas de "convivência com o semiárido" no território de Juazeiro—Bahia. Universidade Federal de Sergipe. Tese de Doutorado em Geografia. (342 p.). São Cristóvão/SE.

CGEE—Centre for Strategic Studies and Management. (2009). Brazil and climate change: Vulnerability, impacts and adaptation. Brasília, DF: CGEE.

Commonwealth Secretariat. (2005). Promising practices and implications for scaling up girls' education. *Report of the UN Girls' Education Initiative South Asia Workshop held in Chandigarh, India, 20–22 September 2004*. Organised by the Commonwealth Secretariat in partnership with UNICEF India.

Dagnino, R. P., Brandão, F. C., & Novaes, H. T. N. (2004). Sobre o marco analítico-conceitual da tecnologia social. Tecnologia Social: uma estratégia para o desenvolvimento. Lassance, A. and Pedreira, J. (Org.). Rio de Janeiro: Fundação Banco do Brasil.

Dagnino, R. P. (1976). *Tecnologia apropriada: Uma alternativa? Master project—Departamento de Economia*. Universidade de Brasília.

FAO—Food and Agriculture Organization (2009). *Climate change and bionergy challenges for food and agriculture*. High level expert forum—How to feed the world in 2050. Rome.

Fernández, L., Lumbreras, J., Andrade, J. C., & Bogo, J. M. (2012). Exploring co-benefits of clean development mechanism projects. *The International Journal of Climate Change: Impacts and Responses, 3*, 121–142.

Fernández, L., Ventura, A. C., Andrade, J. C. S., & Lumbreras, J. (2016). Synergies and trade-offs between climate change mitigation and adaptation strategies: Lessons from social technologies in the semi-arid region of Bahia, Brazil, Latin America. *Latin American Journal Management for Sustainable Development, 3*, 1–18.

Franzel, S., Denning, G. L., Lillesø, J. P. B., & Mercado, A. R. (2004). Scaling up the impact of agroforestry: Lessons from three sites in Africa and Asia. *Agroforestry Systems, 61*, 329–344.

Gillespie, S. (2004). Scaling up community-driven development: a synthesis of experience. In: *Food consumption and nutrition division discussion paper no. 181*. International Food Policy Research Institute.

Ilha, M. S. O., & Ribeiro, M. F. (2012). Adoption of technology by the low-income population segment: The low-cost hot water heater case. *Habitat International, 36*, 185–191.

IPCC—International Panel on Climate Change (2007). Contribution of working Group III to the fourth assessment report of the intergovernmental panel on climate change. In: Metz, B., Davidson, O. R., Bosch, P. R., Dave, R., & Meyer, L. A. (Eds.). Cambridge University Press, Cambridge, United Kingdom and New York, NY, USA.

ITS—Instituto de Tecnologia Social. (2007). Tecnologia Social. Série conhecimento e cidadania. Availabel at http://www.itsbrasil.org.br/publicacoes. Access in July 19, 2010.

Laukkonen, J., Blanco, P. K., Lenhart, J., Kiener, M., Cavric, B., & Kinuthia-Njenga, C. (2009). Combining climate change adaptation and mitigation measures at the local level. *Habitat International, 33*, 287–292.

Leal, I. R., Silva, J. M. C., Tabarelli, M., & Lacher, T. E., Jr. (2005). Changing the course of biodiversity conservation in the caatinga of Northeastern Brazil. *Conservation Biology, 19*(3), 701–706.

Lindoso, D. P., Rocha, J. D., Debortoli, N., Parente, I. C. I., Eiró, F., Bursztyn, M., & Rodrigues Filho, S. (2011). Climate change and vulnerability to drought in the semi-arid: The case of smallholder farmers in the Brazilian Northeast. In: Seroa da Motta, R., et al. (Ed.), *Climate Change in Brazil: Economic, social and regulatory aspects* (pp. 235–256). Brasília: Ipea.

Luyet, V., Schlaepfer, R., Parlange, M. B., & Buttler, A. (2012). A framework to implement stakeholder participation in environmental projects. *Journal of Environemental Management, 111*, 213–219.

Marengo, J. A., & Bernasconi, M. (2015). Regional differences in aridity/drought conditions over Northeast Brazil: Present state and future projections. *Climatic Change, 129*(1–2), 103–115.

Marengo, J. A., Torres, R. R., & Alves, L. M. (2017). Drought in Northeast Brazil—Past, present, and future. *Theoretical and Applied Climatology, 129*(3–4), 1189–1200.

Miranda, I., Lopez, M., & Soares, M. C. C. (2011). Social technology network: Paths for sustainability. *Innovation and Development, 1*(1).

NACIT—Núcleo de Política e Administração em Ciência e Tecnologia (2010). Relatório de pesquisa do projeto Mapeamento da Inovação no Estado da Bahia: subprojeto Mapeamento e Caracterização das Tecnologias Sociais. Universidade Federal da Bahia. Salvador.

Narváez, J. (1996). Tecnologías apropiadas para el desarrollo sostenible. Perú: ITACAB.

Nobre, C. A., Oyama, M. D., Oliveira, G. S., Marengo, J. A., & Salati, E. (2004). Impact of climate change scenarios for 2100 on the biomes of South América. In: *Proceedings of first international CLIVAR conference*, Baltimore, USA, June 21–25.

Nyong, A., Adesina, F., & Elasha, B. O. (2007). The value of indigenous knowledge in climate change mitigation and adaptation strategies in the African Sahel. *Mitigation and Adaptation Strategies for Global Change, 12*, 787–797. https://doi.org/10.1007/s11027-007-9099-0

Polhmann, A. (2011). Local climate change governance. In: Engels, Anita. (Ed.), *Global transformations towards a low carbon society*, Vol. 5 (Working Paper Series), Hamburg: University of Hamburg/KlimaCampus.

RTS—Rede de Tecnologia Social (Org.) (2010). Tecnologia Social e Desenvolvimento Sustentável: contribuições da RTS para a formulação de uma política de Estado de Ciência, Tecnologia e Inovação. Brasília/DF: Secretaria Executiva da Rede de Tecnologia Social (RTS).

Santos, J. E. (2011). Estratégias de convivência para a conservação dos recursos naturais e mitigação dos efeitos da desertificação no semiárido. Desertificação e mudanças climáticas no semiárido brasileiro. Ministério de Ciência e Tecnologia—MCT (Org.). Campina Grande, PB: Insa-PB.

Santos, A. S. (2008). Social and environmental vulnerabilities in the face of climate change for the semi-arid area of Bahia–Brazil. In K. R. Aryal & Z. Gadema (Eds.), *Climate change and disaster impact reduction*. Northumbria University.

Schumacher, E. F. (1973). *Small is beautiful: Economics as if people mattered*. Harper & Row. ISBN-10: 0061361224. ISBN-13: 9780061361227.

Semiárido Show. (2011). Projeto incentiva a produção sustentável de prequenas propriedades rurais. Published in August 25, 2011. Available at http://semi-aridoshow2011.blogspot.com.br/2011/08/projeto-incentiva-producao-sustentavel.html. Access in April 10, 2012.

Stringer, L. C., Dyer, J. C., Reed, M. S., Dougill, A. J., Twyman, C., & Mkwambisi, D. (2009). Adaptations to climate change, drought and desertification: Local insights to enhance policy in southern Africa. *The Environmental Science & Policy, 12*, 748–765.

Thomas, H. E. (2009). Tecnologias para Inclusão Social e Políticas Públicas na América Latina. In: Rede de Tecnologia Social (RTS), Tecnologias Sociais: caminhos para a sustentabilidade. Brasília/DF, s.n,.

Toni, F., Holanda Jr., E. (2008). The effects of land tenure on vulnerability to droughts in Northeastern Brazil. In: *Global environmental change*. Vol. 18, Issue 4 (Local evidence on vulnerabilities and adaptations to global environmental change), October.

UNISDR—United Nations Office for Disaster Risk Reduction. (2010). *My city is getting ready*. Campaign Network and Global Alliance 2010. Available at https://www.unisdr.org/files/14043_campaignkit1.pdf. Access in March 30, 2018

UNISDR—United Nations Office for Disaster Risk Reduction. (2012). *How to make cities more resilient. A handbook for local government leaders*. Geneva, March 2012. Available at https://www.unisdr.org/files/26462_handbookfinalonlineversion.pdf. Access in April 01, 2018

Ventura, A. C., Fernandez, L., Trujillo, R., Andrade, J. C. S. (2011). Potencial das Tecnologias Sociais para o Enfrentamento das Mudanças Climáticas e para a Promoção Desenvolvimento Humano: Um Olhar sobre o Semiárido Baiano. Bahia Análise & Dados, 21(4), 915–931

Ventura, A. C., Fernández, L., Andrade, J. C. S., Lumbreras, J. (2012). The human side of social technology for climate change mitigation and human development: The case of "Fogões Ecológicos" in Brazil. In: *The international journal of environment and sustainable development, Vol. 11. Special issue the human side of environmental management in Brazilian companies* (pp. 375–393).

Ventura, A. C., Andrade, J. C. S., & Fernandéz, L. (2014). Tecnologias Sociais de Convivência com o Semiárido como Estratégia de Mitigação/Adaptação às Mudanças Climáticas no Brasil. *Astrolabio (Buenos Aires), 12*, 43–72.

Wilbanks, T. J. (2003). Integrating climate change and sustainable development in a place-based context. *Climate Policy, 3*, 147–154.

MIX
Papier aus verantwortungsvollen Quellen
Paper from responsible sources
FSC® C105338

If you have any concerns about our products,
you can contact us on
ProductSafety@springernature.com

In case Publisher is established outside the EU,
the EU authorized representative is:
Springer Nature Customer Service Center GmbH
Europaplatz 3, 69115 Heidelberg, Germany

Printed by Libri Plureos GmbH
in Hamburg, Germany